Optical Methods for Managing the Diabetic Foot

This book discusses optical technologies for Diabetic Foot management. It combines the current medical literature review with an overview of the technology and physics behind it. Thus, it is a single-source introduction to the topic. It can also be used as a reference source and practical guide for the use of technology. The particular focus is on low-cost technologies, including hyperspectral imaging, thermography, and endogenous bacterial fluorescence. Moving diagnostic modalities closer to the patient (e.g., primary care) allows the disease to be detected at an earlier stage, thus improving outcomes. However, while some optical technologies are available commercially, they have not received wide clinical adoption due to gaps in knowledge translation to mainstream medicine. This book aims to narrow this gap with practical illustrations. The book will be of interest to a broad range of healthcare professionals, clinical researchers, engineers, and decision-makers, who are dealing with complications of diabetes.

Key Features:

- Reviews the current state of technologies.
- Provides a practical guide with practical considerations and illustrations.
- Supplies a 360-degree view of the combination of clinical information with a technology background and primers on physics and engineering.

Optical Methods for Managing the Diabetic Foot

Gennadi Saiko
Robert L. Bartlett
Jose L. Ramirez-GarciaLuna

CRC Press is an imprint of the
Taylor & Francis Group, an **informa** business

Front cover image: Billion Photos/Shutterstock
First edition published 2024
by CRC Press
2385 NW Executive Center Drive, Suite 320, Boca Raton FL 33431

and by CRC Press
4 Park Square, Milton Park, Abingdon, Oxon, OX14 4RN
CRC Press is an imprint of Taylor & Francis Group, LLC

© 2024 Gennadi Saiko, Robert L. Bartlett and Jose L. Ramirez-GarciaLuna

This book contains information obtained from authentic and highly regarded sources. While all reasonable efforts have been made to publish reliable data and information, neither the author[s] nor the publisher can accept any legal responsibility or liability for any errors or omissions that may be made. The publishers wish to make clear that any views or opinions expressed in this book by individual editors, authors or contributors are personal to them and do not necessarily reflect the views/opinions of the publishers. The information or guidance contained in this book is intended for use by medical, scientific or health-care professionals and is provided strictly as a supplement to the medical or other professional's own judgement, their knowledge of the patient's medical history, relevant manufacturer's instructions and the appropriate best practice guidelines. Because of the rapid advances in medical science, any information or advice on dosages, procedures or diagnoses should be independently verified. The reader is strongly urged to consult the relevant national drug formulary and the drug companies' and device or material manufacturers' printed instructions, and their websites, before administering or utilizing any of the drugs, devices or materials mentioned in this book. This book does not indicate whether a particular treatment is appropriate or suitable for a particular individual. Ultimately it is the sole responsibility of the medical professional to make his or her own professional judgements, so as to advise and treat patients appropriately. The authors and publishers have also attempted to trace the copyright holders of all material reproduced in this publication and apologize to copyright holders if permission to publish in this form has not been obtained. If any copyright material has not been acknowledged please write and let us know so we may rectify in any future reprint.

Except as permitted under U.S. Copyright Law, no part of this book may be reprinted, reproduced, transmitted, or utilized in any form by any electronic, mechanical, or other means, now known or hereafter invented, including photocopying, microfilming, and recording, or in any information storage or retrieval system, without written permission from the publishers.

For permission to photocopy or use material electronically from this work, access www.copyright.com or contact the Copyright Clearance Center, Inc. (CCC), 222 Rosewood Drive, Danvers, MA 01923, 978-750-8400. For works that are not available on CCC please contact mpkbookspermissions@tandf.co.uk

Trademark notice: Product or corporate names may be trademarks or registered trademarks and are used only for identification and explanation without intent to infringe.

Library of Congress Cataloging-in-Publication Data

Names: Saiko, Gennadi, author. | Bartlett, Robert L. (Robert Laing), 1954- author. | Ramirez-GarciaLuna, Jose L., author.
Title: Optical methods for managing the diabetic foot / Gennadi Saiko, Robert L. Bartlett, and Jose L. Ramirez-GarciaLuna.
Description: First edition. | Boca Raton, FL : CRC Press, 2024. | Includes bibliographical references and index.
Identifiers: LCCN 2023031946 | ISBN 9781032463513 (hardback) | ISBN 9781032469584 (paperback) | ISBN 9781003383956 (ebook)
Subjects: MESH: Diabetic Foot–diagnosis | Diabetic Foot--therapy | Optical Devices | Optical Phenomena
Classification: LCC RC918.D53 | NLM WK 835 | DDC 616.4/62–dc23/eng/20231004
LC record available at https://lccn.loc.gov/2023031946

ISBN: 978-1-032-46351-3 (hbk)
ISBN: 978-1-032-46958-4 (pbk)
ISBN: 978-1-003-38395-6 (ebk)

DOI: 10.1201/9781003383956

Typeset in Minion Pro
by Deanta Global Publishing Services, Chennai, India

Contents

About the Authors, vi

Foreword, ix

CHAPTER 1	Introduction	1
CHAPTER 2	Pathophysiology of the Diabetic Foot	7
CHAPTER 3	Current Diagnostic Methods	13
CHAPTER 4	Anatomic Imaging	28
CHAPTER 5	Optical Diagnostic Techniques	48
CHAPTER 6	Therapeutic Approaches	102
CHAPTER 7	Future Directions	122

APPENDIX A: SKIN AND WOUND MORPHOLOGY, 135

APPENDIX B: BIO OPTICS PRIMER, 149

APPENDIX C: PHYSIOLOGICAL IMAGING: DESIGN CONSIDERATIONS, 182

APPENDIX D: ARTIFICIAL INTELLIGENCE, 193

INDEX, 201

About the Authors

Gennadi Saiko, PhD

Dr. Gennadi Saiko is a distinguished scientist and dynamic entrepreneur revolutionizing MedTech. With a career spanning prestigious research institutes, Fortune 500 companies, and prominent startups, he is a true pioneer in optical healthcare technologies.

Holding an MSc and PhD in Physics from the esteemed Moscow Institute of Physics and Technology, Dr. Saiko seamlessly combines academic brilliance with business acumen. His primary focus is transforming wound care, cardiology, emergency, and critical care through innovative optical technologies.

Dr. Saiko developed several novel imaging modalities for gastroenteroscopy, cardiology, and wound care. One of his innovations (a handheld multimodal imaging system for wound care) was successfully commercialized as the Swift Ray 1 by Swift Medical Inc.

As a leading industry expert, Dr. Saiko consults multiple MedTech companies on technology development and regulatory affairs.

Dr. Saiko is an esteemed author who has published over 30 peer-reviewed journal articles, a book, and seven book chapters. He presented his leading-edge findings at more than 30 international conferences and serves on the Executive Committee of the International Society on Oxygen Transport to Tissue (ISOTT).

Dr. Saiko's impact extends to the scientific community as a reviewer for over 20 esteemed journals and a member of scientific publication boards. Currently an Adjunct Professor in the Department of Physics at Toronto

Metropolitan University (formerly Ryerson University), he is dedicated to nurturing the next generation of scientific minds.

Gennadi Saiko

LinkedIn: https://www.linkedin.com/in/GennadiSaiko

Robert L Bartlett, MD, CPE
As a medical innovator, Dr. Bartlett has created numerous companies solving practical problems: Data Pioneers (SaaS-based quality management), Biomaticsco (hyperspectral wound imaging) –founder, Clinical-Connections (SaaS-based medical communications), MedaNextco (SaaS-based patient-centered care plans) – founder with Microsoft alumni, and iCertus Health (SaaS-based clinical decision support).

Dr Bartlett holds multiple specialty certifications. These include Certified Physician Executive (CPE), Emergency Medicine, Hyperbaric Medicine, and a Masters in Wound Care.

Notable professional accomplishments include the following: President, Council for Medical Education and Testing, Chief designer for the first physician-specific wound care certification exam, Chief Designer and Editor for the American College of Hyperbaric Medicine physician certification exam, Eric Kindwall award (American College of Hyperbaric Medicine), and co-founder, Academy of Physicians in Wound Healing.

He has served as the Chief Medical Officer for two of the largest U.S. wound care companies, National Healing and RestorixHealth, collectively managing 3,000 providers across 500 hospitals. Educationally, he has trained more than 6,000 providers in wound care and hyperbaric medicine. As the Chief Medical Officer for Swift Medical he had the privilege to work with the co-authors of this book developing advanced wound imaging and AI analytics.

Bob Bartlett MD

LinkedIn: https://www.linkedin.com/in/BobBartlettMD/

Jose L. Ramirez-GarciaLuna, MD, MSc, PhD

Dr. Jose L. Ramirez-GarciaLuna is a highly accomplished medical professional with a diverse educational background and extensive experience in clinical practice, academia, and the medical technology industry. He earned his MD and an MSc in Epidemiology and Biostatistics from Universidad Autonoma de San Luis Potosi in Mexico and is certified as an emergency and critical care physician.

Driven by his passion for science, Dr. Ramirez-GarciaLuna earned a PhD in Experimental Surgery and a Certificate in Surgical Innovation at McGill University in Canada in 2020. His research focuses on biomaterials for wound healing, translational research, the immunology of wound healing, innovation, and e-Health. He is considered a pioneer in the clinical translation of hyperspectral imaging, machine learning, and artificial intelligence to address surgical challenges, with a particular emphasis on wound healing and infection detection. His dedication and expertise have resulted in the publication of over 60 peer-reviewed articles.

Dr. Ramirez-GarciaLuna's contributions have been recognized through various accolades, including the Rising Star Award from McGill University, the Canadian Institutes for Health Research Excellency Award, and the distinction of being a Fellow of the Mexican National Council for Science and Technology. Currently, Dr Ramirez-GarciaLuna acts as a Medical Technology Consultant for several companies in Canada, the US, and Europe. He also actively contributes to academia as a Lecturer in the Division of Experimental Surgery at McGill University and as an Adjunct Professor in the Faculty of Sciences at Universidad Autonoma de San Luis Potosi in Mexico.

Foreword

Although the 'splitting of the atom' was a marvelous milestone in physics, Isaac Newton's 'splitting of light' with a prism was more profound. It made the invisible components of white light 'visible' and marked the birth of spectroscopy. It also shaped James Maxwell's revelation of the previously invisible electromagnetic spectrum, which is foundational to all forms of imaging.

In 1676, Antonie van Leeuwenhoek using a single-lens microscope, made the previously 'invisible world' of microorganisms visible and ushered in a new era of medicine. The modern field of biophotonics is transforming invisible physiologic phenomena into visible images at the bedside. In short, it is making the 'invisible' visible.

The field of biooptics and biophotonics is a convergence of numerous disciplines using spectroscopy, fluorescence, and digital optics. This book is designed to meet the needs of clinical and non-clinical readers. The principal focus is the emerging role of biophotonics in the management of diabetic ulcers. For readers new to spectroscopy and fluorescence, it is recommended to begin with Appendix B, entitled 'Bio Optics Primer.' For readers with a technical background, we suggest beginning with Appendix A entitled 'Skin and Wound Morphology.' For all readers, Appendix C, 'Physiologic Imaging: Design Considerations,' provides practical insights for overcoming sources of image interference and improving imaging methods and interpretation.

Gennadi Saiko
Jose Ramirez-GarciaLuna
Robert L. Bartlett
June 30, 2023

CHAPTER 1

Introduction

SHIFTS IN DIET, LIFESTYLE, and exercise drive the growth of diabetes in all countries. In 2013, 37.3 million Americans, or 11.3% of the population, had diabetes. For senior citizens older than 65 years, the prevalence is 27%. The U.S. Center for Disease Control estimates for prediabetes is 38%,which continues to rise annually.

Diabetic Foot Ulceration (DFU) is a major complication of diabetes. The annual prevalence of Diabetic Foot Ulcers is estimated to be 4–10%, and the lifetime risk of developing these ulcers in people with diabetes is estimated to be anywhere from 15% to 25%. Foot ulceration increases the risk of lower-limb amputation, one of the most debilitating complications of diabetes. If left untreated, Diabetic Foot Ulcers may become infected and lead to the total or partial amputation of the affected limb. Around 50% of non-traumatic lower-limb amputations are attributable to diabetic ulcers.[1]

If the integrity of the skin has been breached (abrasions, lacerations, punctures, or avulsions), the wound-healing processes begin. Normally, they occur in a sequence: homeostasis, inflammation, proliferation, and maturation. However, the normal wound-healing cycle can be disrupted. Therefore, a chronic wound can be defined as physiologically impaired due to a disruption of the wound-healing process caused by impaired angiogenesis, innervation, or cellular migration, among other reasons. The wound is considered chronic if it does not heal within three (or four, by some sources) weeks.

Diabetic Foot Ulcers are considered chronic wounds. The Wound Healing Society classifies chronic wounds into four major categories:

DOI: 10.1201/9781003383956-1

pressure ulcers, Diabetic Foot Ulcers, venous ulcers, and arterial insufficiency ulcers. As such, DFUs are one of the most prevalent types of chronic wounds.

The DFUs can be divided into neuropathic, ischemic, and neuroischemic. The most prevalent of the three is the neuroischemic DFU, which comprises approximately 50% of such ulcerations.

Diabetic Foot Ulcers, in many cases, start from a benign condition. As a patient has sensory diabetic neuropathy, they do not feel cut or sore. However, with time the cut could get worse and become infected. In addition, contact with soil increases the probability of contamination and infection manifold. Similarly, the foot could be aligned improperly due to neuropathy, creating too much pressure on one part of the foot.

For people with diabetes, many common foot problems can cause DFU. They include athlete's foot, fungal infection of nails, calluses, corns, blisters, bunions, dry skin, hammertoes, ingrown toenails, and plantar warts.

Even when ulcers have healed, clinicians should view Diabetic Foot Ulcers as a lifelong relapsing condition that requires monitoring to minimize the annual recurrence frequency. Sixty percent of patients with a history of ulceration develop another ulcer within one year of wound healing.[2] Diabetes increases the incidence of foot ulcer admissions by 11-fold, accounting for more than 80% of all amputations and increasing hospital costs more than 10-fold over the five years.[3]

To prevent dreadful complications of Diabetic Foot Ulcers, the condition should be managed appropriately.

Diabetic Foot Ulcer management aims to help prevent serious complications and improve the quality of life for individuals with diabetes. DFU management is a set of procedures and treatments to treat and heal foot ulcers in individuals with diabetes. Diabetic Foot Ulcer management consists of multiple steps, including:

- Clean the wound thoroughly and remove any dead tissue to promote healing;
- Control the infection by prescribing antibiotics or other medications if needed;
- Off-load pressure from the wound using special shoes, casts, or other devices;
- Provide wound dressings to protect the wound and promote healing;

- Control blood glucose levels to optimize the healing process;
- Monitor the wound regularly to track its progress and identify any new complications;
- Educate individuals with diabetes about the importance of proper wound care and on how to prevent new ulcers from forming.

Diabetic Foot Ulcer management requires a multi-disciplinary approach involving the coordinated efforts of a team of healthcare professionals, including a doctor, nurse, physical therapist, and podiatrist.

The important concept in any clinical practice is a clinical workflow, which refers to an organized and repeatable process for assessing a patient, planning and executing treatment, and documenting all care provided to the patient.

Efficient wound care workflows typically involve the following steps:

- Comprehensive patient assessment involves thoroughly examining the patient's medical history, current condition, and wound characteristics. This information is used to inform the diagnosis and treatment plan.
- Wound diagnosis: once the patient assessment is complete, the wound is diagnosed based on its type, size, location, and severity. This information is used to select the appropriate treatment option.
- Treatment planning: based on the wound diagnosis, a treatment plan outlines the steps required to manage the wound effectively, which may involve dressings, antibiotics, debridement, or other wound care interventions.
- Treatment implementation: once the treatment plan has been developed, the wound care team implements the plan and monitors the patient's progress over time.

Optical methods play an essential role in DFU management. They are used across the whole clinical workflow. Based on their clinical utility and relevance to a particular task of the clinical workflow, optical technologies in DFU management can be split into three large groups: anatomical imaging, physiological imaging, and therapeutic applications.

What is anatomical and physiological imaging?

By anatomical imaging, we refer to the traditional RGB or monochrome imaging. In a typical anatomical imaging scenario, the nurse takes an image of the wound using a DSLR camera or a smartphone. This image is then either stored in Electronic Health Record(EHR) or processed, and some geometrical information (e.g., wound area, length, width, depth) is retrieved automatically, semi-automatically, or manually. Then, this information can be used to track the healing progress in time. Thus, anatomical imaging is particularly helpful for monitoring, including the measurement and documentation of wounds. Therefore, anatomical imaging and other monitoring techniques will be the focus of Chapter 4 (Anatomic Imaging).

Physiological imaging refers to using imaging technologies to extract physiological or pathophysiological parameters. In this case, the imaging technology is augmented by some extra functionality. For example, it can be laser Doppler, hyperspectral, or fluorescence imaging. This group of technologies is typically used for screening, triaging, and diagnostic purposes. Therefore, optical diagnostic techniques and physiological imaging will be the focus of Chapter 5 (Optical Diagnostic Techniques).

The book is organized as follows:

In Chapter 2, we will overview the pathophysiology of diabetic foot. In particular, we will discuss polyneuropathy and vasculopathy, which include macrovascular and microvascular diseases.

Chapter 3 will overview current diagnostic methods routinely used in the clinic.

Chapter 4 will overview the monitoring techniques with a primary focus on anatomical imaging methods. Here we included planimetric and stereometric applications, which can be used for wound margin delineation and assessment of wound area and its length, width, and depth. In addition, we included pure image processing techniques, which classify wound (epithelial vs. granulation vs. eschar vs. slough) or periwound (e.g., erythema[4]) tissue types.

In Chapter 5, we will overview optical diagnostic methods. Nowadays, most diagnostic optical technologies used in wound care are imaging-based. Imaging modalities have some inherited advantages that are particularly relevant for healthcare. Firstly, in most cases, they are noncontact (with ultrasound-based technologies, which require contact with skin, as a notable exception). It is beneficial to prevent bacterial contamination. It also benefits a broad range of patients (including people with diabetes) with sensitive skin. Secondly, it allows extracting information from some areas, which allows immediate comparison between different skin areas (self-control). While some contact modalities (like Photoplethysmographic or PPG devices) may still exist in labs, most incoming technologies are imaging-based. Thus, the focus of this book will be primarily on imaging technologies.

In Chapter 5, we will focus on physiological imaging techniques, which have been translated into the clinic but have yet to receive widespread clinical adoption. These methods focus on extracting physiological parameters used primarily for screening and diagnostic purposes. Efficient screening modalities allow earlier interventions in a patient population that presents clinically with late-stage complications, significant morbidity, and mortality risk.

Thus, we have included optical modalities based on their relevance to primary diabetic foot care management. Consequently, we have excluded several modalities (e.g., Optical Coherence Tomography (OCT), photoacoustic imaging, Raman spectroscopy, microscopy, and video capillaroscopy) as having a limited clinical impact on diabetic wound care and primary care as management tools. These modalities are helpful research tools but are limited in clinical care due to their expense, complexity, and time requirements for screening measurement.

Chapter 6 will discuss therapeutic applications of optical technologies in DFU management. Light-based technologies can be used for traditional applications in wound care as surgery, debridement, and antiseptic. In addition, in the past 50 years, a significant body of knowledge has been accumulated on the healing effects of light. This technology group can be called Low-Level Laser Therapy (LLLT). Thus, in addition to traditional light applications, we will also discuss LLLT-based technologies.

In Chapter 7, we will discuss new frontiers in DFU management. With the rapid penetration of Artificial Intelligence (AI) in all aspects of daily life, we cannot ignore its influence on wound management. Thus, this chapter will focus on applications of AI and other trends in wound care and DFU management.

We tried to make the book self-contained and included as much supplementary information as possible. We moved all background and supplemental information from the primary flow to avoid constantly switching the topic from physics and basic science to clinical utility. As such, we created four appendixes. In Appendix A, we provide basic information on skin and wound morphology. In Appendix B, we provide basic information on tissue optics. Appendix C includes basic information on image collection and processing techniques. Finally, Appendix D provides basic information on AI and machine learning approaches relevant to wound care and DFU management.

NOTES

1. A Raghav, ZA Khan, RK Labala, et al. "Financial burden of diabetic foot ulcers to world: A progressive topic to discuss always", *Therap Adv Endocrinol Metab*, 9:29–31 (2018).
2. BJ Petersen, SA Bus, GM Rothenberg, et al. "Recurrence rates suggest delayed identification of plantar ulceration for patients in diabetic foot remission", *BMJ Open Diabetes Res Care*, 8(1):1–8 (2020).
3. DG Armstrong, AJM Boulton, SA Bus, "Diabetic foot ulcers and their recurrence", *N Engl J Med*, 376:2367–2375 (2017).
4. M Ptakh, G Saiko, "Developing a robust estimator for remote optical erythema detection", In Proceedings of the 14th Int Joint Conf on Biomed Eng Sys Techn (BIOSTEC 2021) – Vol 2: BIOIMAGING, 115–119 (2021).

CHAPTER 2

Pathophysiology of the Diabetic Foot

WOUNDS THAT FAIL TO progress through the normal stages of healing and are still open at one month are considered 'chronic wounds'.[1] Failure to heal is typically multifactorial, with several different disease processes (comorbidities) interacting. Arterial disease, diabetes, obesity, immune deficiencies, and malnutrition are common comorbidities that delay healing.[2] Diabetic Foot Ulcer (DFU) is one of the most common types of chronic wounds. The pathophysiology that leads to foot ulceration, impaired healing, and limb loss is a triad of polyneuropathy, vasculopathy, and immune defects. The interplay of these defects is greater than the sum of each acting alone. In brief:

- Neuropathy impairs the protective sensation of the foot, and this leads to daily micro-trauma of the bones and soft tissues;
- Obesity increases the working load of the foot and the magnitude of daily trauma to the insensate foot;
- Vascular disease impairs the delivery of blood to the extremity impairing the recovery from daily micro-trauma bacterial invasion;
- Immune deficiencies due to high blood sugars impair the response to infection.

2.1 POLYNEUROPATHY

Polyneuropathy is a common complication of diabetes and encompasses several neuropathic syndromes. Neuropathy can be symmetric or asymmetric; the symptoms can be attributed to impaired sensation, motor function, or a combination. Diabetic neuropathy may affect various combinations of sensory, motor, and autonomic neuropathy. The clinical presentation is varied (see Table 2.1).

Distal symmetric neuropathy with sensory impairment is the most common form affecting 30% of all people with diabetes.[3] Because the longest nerves are the first to be affected, the earliest symptoms appear in the feet. It is important to understand that many patients with distal sensory neuropathy may not be aware of early sensory impairment and motor loss. These individuals' first symptom may be a relatively painless, nonhealing foot ulcer.

The astute clinician should take care to avoid 'anchoring bias.' Diabetic neuropathy is a diagnosis of exclusion. Nondiabetic, treatable neuropathies may be present in patients with diabetes (see Table 2.2).

Diabetic nerve damage is multifactorial. The damage is attributed to endothelial dysfunction affecting the vasa nervorum, which provides blood to larger nerves, oxidative stress from hyperglycemia, impaired antioxidant production, and advanced glycation end products. Epidemiology studies indicate that neuropathy is more likely with increased age, duration of diabetes, poor glycemic control, and vascular disease.

A thorough foot examination is important to detect the disease early. Screening for peripheral neuropathy and Peripheral Arterial Disease (PAD) can help identify patients at risk for foot ulcers. A history of ulcers or amputations and poor glycemic control increases the risk of future ulceration.

TABLE 2.1 Diabetic neuropathic presentations

- Chronic sensory neuropathy
- Symmetric distal motor neuropathy
- Proximal motor neuropathy
- Focal vascular neuropathy
- Truncal radiculopathy
- Acute painful neuropathy of poor glycemic control
- Acute painful neuropathy of rapid glycemic control
- Cranial mono-neuropathy (CN III)

TABLE 2.2 Differential diagnosis of distal symmetric neuropathy

Drugs	Metabolic
• Alcohol	• Diabetes
• Chemotherapy	• Uremia
• Nitrofurantoin	• Myxedema
• Isoniazid	• Porphyria
• Colchicine	• Paraproteinemia
• Amiodarone	• Vitamin deficiency
• Dapsone	• Mercury
	• Arsenic
	• Neoplastic syndromes
	• Bronchial carcinoma
	• Gastric carcinoma
	• Lymphoma
Infectious	Inflammatory
• Leprosy	• Guillain Barré Syndrome
• Lyme borreliosis	• Polyarteritis nodosa inflammatory demyelinating polyneuropathy
• HIV	
	Genetic
	• Hereditary sensory neuropathies
	• Charcot-Marie-Tooth disease

2.1.1 Neuropathy Consequences

Impaired sensory functions are associated with painful symptoms and loss of protective sensation. Impaired motor function leads to atrophy and loss of muscle tone. Because the classic form of diabetic neuropathy is distal, the foot's intrinsic muscles are affected more than the proximal muscles. When this occurs, the tendons from the large proximal muscles are no longer balanced by normal muscle tone from the intrinsic muscles of the feet. This leads to numerous foot deformities such as 'hammertoes' and buckling of the metatarsal heads –'clawed foot.' The deformed foot is less efficient and subject to mechanical stress and subsequent 'painless' soft tissue breakdown due to loss of protective sensation.

Impaired autonomic function leads to altered vascular tone and malperfusion of the foot. Additionally, the autonomic system governs oil and sweat glands. Neuropathy is associated with the loss of these glands, leading to dry, cracked skin prone to infection.

2.2 VASCULOPATHY

Diabetes-associated vasculopathy occurs at a macrovascular and microvascular level. The macrovascular disease affects large blood vessels. The microvascular disease affects very small vessels, less than 300 micrometers.

Some of these vessels contain smooth muscle, which regulates blood flow to the capillary beds. The terminal vessels of the microcirculation are the capillaries that lack smooth muscles.

2.2.1 Macrovascular Disease

Macrovascular disease can be atherosclerotic occlusive disease or non-occlusive medial calcinosis, which is a loss of arterial wall elasticity. The proatherogenic changes associated with diabetes include increases in vascular inflammation and derangements in the cellular components of the vasculature and alterations in blood cells and hemostatic factors. Patients with diabetes have a fourfold increase in the prevalence of atherosclerosis and are more likely to have an accelerated course. They are also more sensitive to moderate changes in perfusion compared to patients without diabetes. Patients with diabetes have a propensity for diffuse disease below the knee with minimal disease in the pedal arteries – specifically the anterior tibial artery, the posterior tibial artery, and the peroneal artery. In contrast, nondiabetic patients have a more discreet plaque-like disease of the proximal arteries. These anatomical observations are important when planning surgical interventions to improve blood flow.

The most common Peripheral Arterial Disease (PAD) symptom is claudication, defined as pain, cramping, or aching in the calves, thighs, or buttocks, which appears with walking exercise and is relieved by rest. When significant neuropathy is present, these symptoms will be masked despite significant arterial occlusion. Additionally, patients with sedentary lifestyles may not walk far enough to experience classic claudication pain. Collectively, these considerations explain why many patients with diabetes may have severe disease that is clinically not apparent due to the lack of symptoms. For some, a nonhealing wound is the first symptom of severe arterial disease.

As PAD progresses, the pain will occur at rest and may awaken patients during the night. Oftentimes, these patients find relief by dangling their legs over the side of the bed with gravity assisting the arterial flow. In extreme cases, the tissue dies and turns black (dry gangrene) or may become infected (wet gangrene).

2.2.2 Microvascular Disease

Diabetes causes structural and functional changes in microcirculation. The most prominent structural change is the thickening of the basement membrane, which supports the endothelial cells. These microvascular changes are more pronounced in the lower extremities. Changes in the basement membrane affect multiple functions, such as vascular permeability, cellular adhesion, and proliferation, which delay wound healing.

Membrane thickening impairs oxygen diffusion, carbon dioxide, nutrients, and leukocyte migration from the lumen to the interstitial space outside the capillaries. This limits the ability to control infection. The blood vessels become less elastic with thicker basement membranes, limiting vasodilation in response to injury and stress – decreasing the delivery of oxygen, nutrients, and the immune response. It is important to note the thickening is not an occlusive process, as the luminal diameter is unchanged.

The impaired vasodilation of the microcirculation in response to injury is described as *functional ischemia*. This is to say, there is a mismatch of supply to local demand at the capillary level. The loss of appropriate dilation of the metarterioles results in a functional arteriovenous shunting of blood past the tissue beds in need.[4,5] Metarterioles form a precapillary sphincter and govern the blood flow to the capillaries, where the exchange of gas and nutrients occurs.

Diabetic Foot Ulceration is a chronic complication of diabetes associated with neuropathy and/or arteriosclerotic peripheral vascular diseases in the lower limb, alongside microvascular abnormalities.[6,7] There are two physiological theories pertaining to diabetic microvascular disease. The *capillary steal premise* is based on the sympathetic autonomic neuropathy in the lower limb. The loss of appropriate vasoconstrictor control and alteration in blood flow increase arteriovenous shunting and reduce nutritive blood flow.

The *hemodynamic premise* is supported by evidence of hyperglycemic oxidative stress and cell damage, which limit nitric oxide (NO) production.[8,9] Nitric oxide is an important signaling molecule for vasodilation and other cell processes. The downstream consequences of the hemodynamic model are microvascular remodeling, thickening of capillary basement membranes, microvascular sclerosis, and loss of autoregulation and nutritive flow.[10,11,12,13]

NOTES

1. GS Lazarus, DM Cooper, DR Knighton, et al. "Definitions and guidelines for assessment of wounds and evaluation of healing", *Wound Repair Regen*, 2:165–170 (1994).

2. R Waaijman, M de Haart, ML Arts, et al. "Risk factors for plantar foot ulcer recurrence in neuropathic diabetic patients", *Diabetes Care*, 37(6):1697–1705 (2014).
3. A Veves, JM Giurini, FW LoGerfo (editors), *The Diabetic Foot*, 2nd Edition, Humana Press, Totowa, New Jersey, USA, 2006, p. 105.
4. MD Flynn, JE Tooke, "Diabetic neuropathy and the microcirculation", *Diabet Med*, 12(4):298–301 (1995).
5. A Behroozian, JA Beckman, "Microvascular disease increases amputation in patients with peripheral artery disease", *Arterioscler Thromb Vasc Biol*, 40(3):534–540 (2020).
6. AJ Boulton, "The pathogenesis of diabetic foot problems: An overview", *Diabet Med*, 13:S12–16 (1996).
7. MD Flynn, JE Tooke, "Aetiology of diabetic foot ulceration: A role for the microcirculation?", *Diabet Med*, 9:320–329 (1992).
8. HH Parving, GC Viberti, H Keen, et al. "Hemodynamic factors in the genesis of diabetic microangiopathy", *Metabolism*, 32:943–949 (1983).
9. DD Sandeman, AC Shore, JE Tooke, "Relation of skin capillary pressure in patients with insulin-dependent diabetes mellitus to complications and metabolic control", *N Engl J Med*, 327:760–764 (1992).
10. CYL Chao, GLY Cheing, "Microvascular dysfunction in diabetic foot disease and ulceration", *Diabetes Metab Res Rev*, 25:604–614 (2009).
11. L Uccioli, L Mancini, A Giordano, et al. "Lower limb arterio-venous shunts, autonomic neuropathy and diabetic foot", *Diabetes Res Clin Pract*, 16:123–130 (1992).
12. JE Tooke, "Peripheral microvascular disease in diabetes", *Diabetes Res Clin Pract*, 30:S61–S65 (1996).
13. D Lowry, M Saeed, P Narendran, et al. "The difference between the healing and the nonhealing diabetic foot ulcer: A review of the role of the microcirculation", *J Diabetes Sci Technol*, 11:914–923 (2017).

CHAPTER 3

Current Diagnostic Methods

3.1 NEUROPATHY EXAMINATION

The American Diabetes Association 2020 guidelines recommend assessing diabetic peripheral neuropathy starting at the diagnosis of type 2 diabetes, five years after the diagnosis of type 1 diabetes, and at least annually thereafter. Assessment for distal symmetric polyneuropathy should include a careful history and assessment of either temperature or pinprick sensation (small fiber function) and vibration sensation using a 128Hz Tuning Fork (large fiber function). All patients should undergo 10-gram monofilament testing annually to identify whether feet are at risk for ulceration and amputation. Up to 50% of diabetic peripheral neuropathies may be asymptomatic. These patients are at risk for injuries to their insensate feet if the sensory impairment is not recognized early with preventive foot care.

Symptoms vary according to the type of sensory fibers affected. Typically, small fibers are affected early, producing pain and dysesthesia (burning sensation). Large fiber involvement is associated with tingling, numbness, progressive loss of protective sensation, and loss of balance.[1]

The loss of the 'gift of pain' places patients at risk for recurring injury and foot ulceration over the pressure points on the foot. Because of the loss of feedback on pain, it is common for patients to walk on these open ulcers for weeks before they are discovered.[2]

DOI: 10.1201/9781003383956-3

3.1.1 Small Fiber Tests

- Pinprick – Gently touch the skin with a pin or back end of a pin and ask the patient whether it feels sharp or dull. Begin at the forefoot and move proximally. Record the level where sharp/dull discrimination begins.

- Temperature – Use two test tubes, one filled with cold water and the other with warm water. Touch the skin for 2 seconds and ask the patient, 'Hot or cold?' Begin at the forefoot and move proximally. Record the level where hot/cold discrimination begins.

3.1.2 Large Fiber Tests

- Vibration – Use a 128Hz Tuning Fork. The test is considered positive when the patient loses vibratory sensation over the metatarsal heads, whereas the examiner still perceives it when applied to their distal radius at the wrist. Begin using the first dorsal metatarsal–phalangeal head. If that location is positive for neuropathy, assess over the medial and lateral malleoli at the ankle.

- Light touch – The Semmes–Weinstein Monofilament (SWM) examination uses a 10-gram monofilament to perform a 10-point assessment, as illustrated in Figure 3.1. The monofilament is held in place for about 2 seconds. Press the monofilament to the skin so that

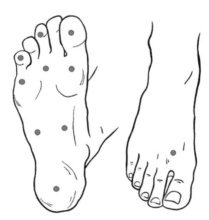

FIGURE 3.1 Ten-point Semmes–Weinstein examination. Nine points are located on the plantar surface, and one is on the dorsal surface. The examination is positive for neuropathy if the patient fails to perceive three or more locations.

it buckles at one of the two times as you say, 'Time one' or 'Time two.' Have the patient identify at which time they were touched. Randomize the sequence of applying the filament throughout the examination. If there is an ulcer, callus, or scar on the foot, the monofilament is applied to an adjacent area.

The SWM examination provides a convenient screen for loss of protective sensation. The inability to feel the 10 grams of force or a 5.07 monofilament applied is associated with clinically significant large fiber neuropathy and a high risk of foot complications. In three studies, the SWM examination identified foot ulceration risk with a sensitivity of 66–99%, a specificity of 34–86%, a positive predictive value of 18–39%, and a negative predictive value of 94–95%.[3,4,5,6]

3.2 MACROVASCULAR ASSESSMENT

Early diagnosis of peripheral arterial disease is important because it signals the need to evaluate coronary, renal, and cerebral arteries. The absence of peripheral pulses on physical examination would support the diagnosis of Peripheral Arterial Disease (PAD). However, patients can have a significant disease before the pulses disappear. The physical examination as a 'test' lacks sensitivity and has poor agreement between examiners. Better studies are available for PAD screening. Commonly used are the Ankle Brachial Index (ABI) and the handheld Doppler. Guidelines frequently cite them as reasonable screening tests as they are inexpensive, use tools readily available in most clinics, and can be performed at the bedside.[7]

3.2.1 Ankle Brachial Index

The ABI is performed by measuring the systolic blood pressure in the ankles (dorsalis pedis and posterior tibia arteries) and arms (brachial artery) using a handheld Doppler and then calculating the ratio. The ABI is measured by placing the patient in a supine position for 5 minutes. Systolic blood pressure is measured in both arms, and the higher value is used as the denominator of the ABI. Systolic blood pressure is then measured in the dorsalis pedis and posterior tibia arteries by placing the cuff just above the ankle. The higher value is the numerator of the ABI in each limb (Table 3.1).

The ABI is a simple, noninvasive, quantitative measurement of the patency of the lower extremity arterial system. Compared to pulse

TABLE 3.1 Diagnostic criteria for PAD using ABI measurements

ABI Value	Interpretation
0.91–1.30*	Normal
0.70–0.90	Mild obstruction
0.40–0.69	Moderate obstruction
<0.40	Severe obstruction

* An ABI value >1.3 is unreliable and suggests poorly compressible arteries at the ankle level due to the presence of medial arterial calcification

palpation, the ABI is much more accurate. It has been validated against contrast arteriography and found to be 95% sensitive and almost 100% specific.[8]

3.2.1.1 ABI Limitations

With normal arteries, the ankle pressure is similar to the brachial pressure producing an ABI of 1. With occlusive atherosclerotic disease, the ankle pressure is lower because the upstream narrowing of the blood vessels produces a downstream drop in pressure at the ankle in proportion to the decreased diameter and the length of the narrowed segment. The ABI is also sensitive to serially located lesions which have a cumulative effect. Where there are multiple lesions located in the iliac, femoral, and popliteal arteries, each will contribute to a decrease in distal pressure. This makes the ABI a good measure of the severity and number of atherosclerotic lesions affecting blood flow to the foot.[9]

However, with diabetes, medial calcinosis that stiffens the arterial walls is a common problem. The stiff arterial walls resist external compression by the blood pressure cuff, which falsely elevates the ABI because blood flow remains detectable by Doppler. Atherosclerosis in patients with diabetes differs in another important way. Anatomically, the lesions predominantly reside in the Below-The-Knee (BTK) arteries, and proximal arteries are relatively disease-free.[10,11,12] This unique distribution of atherosclerotic disease affects the ABI predictive value for PAD. In particular, the ABI can vary greatly depending on which BTK artery is chosen for ankle pressure measurement. An ABI may even be normal if a patent BTK artery is selected for measurement, even when the two other arteries are severely diseased. However, this is exactly what the standard ABI methodology does.[13] It takes the highest available distal pressure from either the dorsal pedal or the posterior tibial artery as the ankle pressure value and creates an inherent bias to underestimate the degree of atherosclerosis.

There is also the error induced by stiffened arteries that has to be considered. Medial calcinosis is not exclusively confined to the leg and may also affect the brachial arteries. This is important as the brachial arteries serve as the 'reference value' and are assumed to be normal. A significant difference (>15%) between the left and the right brachial pressure can be found in 7% of patients with suspected coronary artery disease and higher rates in patients with diabetic foot.[14]

The limitations of ABI screening for patients with diabetes were studied in an extensive series of 187 lower extremities with a diabetic foot ulcer with an intra-arterial angiography and an ABI.[15] The extent of atherosclerosis on angiography was rated by scoring all arterial segments from the aorta to the foot, conforming to the Joint Vascular Societies reporting standard. Arterial calcification was assessed using a 4-level severity scale based on the number and length of calcified arterial segments as seen on a plain X-ray.

A meaningful ABI could only be determined in 123 cases (65.7%); the mean ABI was 0.92. Analysis of the angiographies showed that atherosclerotic lesions had a high predilection for BTK arteries. The correlation between ABI and angiographic PAD was weak (Pearson correlation coefficient $r = 0.487$).

The ABI also has conceptual limitations. The ABI premise asserts that the pressure measured distally is representative of atherosclerotic disease more proximally. This premise is reasonable for the iliac, femoral, and popliteal arteries, which are arranged in a series. Stenosis in these arteries will contribute to a distal pressure drop. However, with diabetes, the affected vessels are the three BTK arteries which are arranged in parallel with each other. Thus, the pressure measured in one artery is independent of the other two arteries because they are not arranged in a series. Consequently, the ABI value greatly depends on which distal pressure is used for the calculation.

Detecting PAD is important because it is a marker for generalized atherosclerosis and cardiovascular disease and predicts wound healing. There is a definite need for a new method that is simple, fast, reproducible, and reliable. This is the future potential of noninvasive optical methods available at the bedside.

3.2.2 Toe Brachial Index

A Toe Brachial Index (TBI) is performed when the ABI is abnormally high due to calcification of the arteries (calcinosis) in the leg which makes them difficult to compress (an abnormally high ABI is >1.3). When this is the case, the TBI is used as an alternative because the digital arteries of the toes do not calcify.

The calculation of the TBI is similar to that of the ABI. It divides the blood pressure of the great toe by the systolic brachial blood pressure. Clinicians can measure toe pressure by placing a small toe cuff around the great toe and attaching a photoplethysmography probe at the pulp of the great toe tip. TBI is considered a better test; however, it is not routinely used as a screen because specialized toe cuffs and photoplethysmography probes are needed, limiting widespread TBI use as a 'simple' screen. For some patients, the great toe is deformed due to 'clawing' (see Chapter 2), edema is present, or wounds are present in locations that prohibit TBI measurements.

3.2.3 Doppler Probe

Simple auscultation with a Doppler ultrasound probe is a more reliable screen than palpation on a physical examination. However, it does require moderate experience. On auscultation, a normal artery should have a triphasic sound. The first phase is the systolic rush of blood; followed by a second phase, the elastic recoil of the arterial wall; and a final small phase, produced by the shock wave of the aortic valve closure. The sound changes to a subdued biphasic pattern as the arterial disease progresses. It becomes a quiet monophasic sound as the arterial disease increases, and with severe disease, no sound is detectable as there is very little blood flow.

The purpose of arterial screening examinations is to determine whether the macrovascular disease is present. If the screening tests are positive, more sophisticated examinations are needed to localize the disease and determine if the patient can be helped by vascular bypass or endovascular interventions such as balloon stents. These 'anatomic' examinations are performed using contrast angiography, Magnetic Resonance Angiography (MRA), or Computerized Tomographic Angiography (CTA), which have various merits that go beyond the scope of this chapter.

3.3 MICROVASCULAR ASSESSMENT

There is a large array of options for evaluating macrocirculation. Studying macrocirculation is easier due to the size of the vessels and the small number of 'named' vessels. In contrast, evaluation of the microcirculation is more difficult because it comprises vessels measured in microns and the vessels number in the millions (capillaries). The technology drivers for macrocirculation studies are the numerous therapeutic options for endovascular procedures to improve large vessel blood flow. However, there are few options for the local control of microcirculation.

From a clinical perspective, it is important to understand that macrocirculation is merely a conduit for delivering oxygen and nutrients. The real 'biology' of exchange occurs at the microcirculatory level. Because diabetes impairs microcirculation, it has been observed that technical correction of macro-flow does not reliably lead to clinical healing of diabetic foot ulcers. An increase in 'macro' flow does not necessarily translate to a rise in micro (nutritive) flow at the capillary level because of significant arteriovenous shunting and the diffusion barrier of a thick basement membrane. Therefore, the objective marker for success should be a measurable change in tissue oxygen.

From a clinical perspective, only three available methods for evaluating microcirculation are available: Skin Perfusion Pressures (SPP), Indocyanine Green (ICG) angiography, and Transcutaneous Oxygen Pressures (TcPO$_2$).

3.3.1 Transcutaneous Oxygen Pressure

TcPO$_2$ measurement is a noninvasive test that assesses the partial pressure of oxygen diffusing through the skin. A 'Clark electrode' measures oxygen using a platinum cathode and a silver anode covered with a thin membrane permeable to oxygen. The electrode is submerged in a small, ringed, normal saline water bath on intact skin. The voltage is applied to the electrodes, and the oxygen molecules are reduced to water:

$$O_2 + 4\,e- + 4\,H+ \rightarrow 2\,H_2O$$

The flow of electrons (current) is directly proportional to the number of oxygen molecules expressed as gas pressure in mmHg.[16]

The water bath serves to capture the transdermal diffusion of oxygen. It also hydrates the skin to improve dermal permeability. The Clarke electrode is heated to 44.5°C to induce local hyperthermia and soften the stratum corneum. The hyperthermia creates maximum capillary vasodilatation for a uniform baseline, removing any 'local' variability from vasoconstrictive signals. The measurements are always peri-wound because the electrodes must be placed on intact skin. TcPO$_2$ interpretation assumes that the values inside the wound are the same or worse than those in the peri-wound tissue. An ideal device would provide measurement in the wound itself; however, this is not possible.

A complete transcutaneous examination of both lower extremities typically requires 1 hour. The study is done with the patient in the supine position, and care must be taken to avoid placing the electrodes over bony prominences, veins, scar tissue, cellulitis, or areas of edema, as these areas are associated with lower values. Performing accurate $TcPO_2$ tests requires skill on the part of the operator.[17] Clinical decisions are based on room air measurements. A normal healthy value in the foot being >50 mmHg and a value <40 mmHg is sufficient hypoxia to impair wound healing. Naturally, the lower the $TcPO_2$ value, the less likely the healing. $TcPO_2$ values <30mmHg indicate that Critical Limb Ischemia (CLI) is present, although other conditions, such as anemia, pulmonary disease, edema, and severe heart failure, should be considered as contributing factors for low values.

Provocative maneuvers are frequently used to improve clinical predictions. One maneuver is to breathe 100% oxygen for 15 minutes and evaluate the change in oxygen pressure. Patients with severe disease will have a 'blunted' response to oxygen with little change. Because transcutaneous testing is frequently used to screen patients for Hyperbaric Oxygen (HBO) therapy, measurements may be performed in the hyperbaric chamber at treatment pressure. HBO therapy involves breathing pure oxygen in a pressurized chamber to increase the oxygen pressure in hypoxic tissues.

Diabetic patients whose $TcPO_2$ values during a hyperbaric session are >200 mmHg are significantly likely to benefit from HBO therapy. Patients whose 'in-chamber' $TcPO_2$ values are <50mmHg are not likely to benefit.[18,19,20,21]

Measurements during HBO therapy appear to be the best predictor for selecting patients who will or will not respond to Hyperbaric Oxygen.[22]

$TcPO_2$ measurement can be used for:

- Assessment of peri-wound oxygenation;
- Screening for peripheral arterial disease;
- Diagnosis of Critical Limb Ischemia;
- Prediction of the likelihood of healing based on oxygen-dependent mechanisms;
- Evaluation and monitoring of compromised flaps;

- Prediction of amputation;
- Selection of optimal amputation level;
- Identification of patients who would benefit from HBO;
- Assessment of response to HBO.

3.3.2 Skin Perfusion Pressure

Skin Perfusion Pressure (SPP) is a noninvasive test that quantitatively analyzes cutaneous blood flow by measuring the blood vessels' 'opening pressure' using a red laser light source. The blood flow to a region is stopped using a blood pressure cuff. The cuff is inflated to a pressure above systolic in the microcirculation (typically 90 mmHg); the pressure is held for 10 seconds or longer until the laser flow indicator reads less than 0.1%.

As the pressure in the blood pressure cuff is gradually reduced, a laser Doppler is used to detect the first movement of red cells in the circulation. Red blood cells scatter the laser beam, and the moment flow resumes, the reflected signal changes frequency (Doppler shift). The pressure associated with the first movement of red cells is the opening pressure for the microcirculation.

Unlike photoplethysmography, which detects the intensity of light scattered from red blood cells, the laser Doppler sensor determines the presence of a Doppler frequency shift generated by the motion of the red cells. A visual recording shows the cuff pressure (x-axis) versus the detected red blood cell motion or perfusion (y-axis).

SPP has several advantages:

- Less time-consuming than $TcPO_2$;
- No calibration is required;
- Anatomical evaluation of vasculosomes;
- Not affected by arterial calcinosis;
- Not affected by moderate edema;
- Not affected by thick skin.

SPP has been studied extensively in both diagnostic and prognostic capacities. It provides quantitative measurement to guide the need for revascularization and the probability of healing lower extremity ischemic ulcers, skin incisions, or amputation sites. The following observations are made in clinical practice:[23,24,25,26]

- <30 mmHg – Critical Limb Ischemia;
- 30–40 mmHg – Impaired wound healing based on perfusion;
- 40–50 mmHg – Abnormal, impaired healing?;
- >50 mmHg – Normal skin perfusion.

SPP has been correlated with other noninvasive wound diagnostic methods to determine if it is more sensitive for some conditions or locations. It was found to be a satisfactory substitute for toe pressure, with high correlation coefficients regardless of whether patients had diabetes (Pearson values of 0.85 and 0.93, respectively).[27] A large clinical study found a high correlation between SPP and ankle or toe blood pressure or $TcPO_2$ (respective coefficients of 0.75, 0.85, and 0.62).[28] When the healing of 94 limbs with gangrene or ulcers was predicted using a receiver operating curve and a cut-off value of 40 mmHg, the sensitivity and specificity obtained using SPP were 72% and 88%, respectively.

3.3.3 Fluorescence Imaging

Exogenous fluorescent imaging provides a convenient method of visualizing microcirculation using Indocyanine Green (ICG). Developed for near-infrared (NIR) photography by the Kodak Research Laboratories in 1955, ICG was approved for clinical use in 1957 and has a well-established safety profile.

The methodology for imaging is a straightforward fluorescent process in which the fluorophore (ICG) is excited by a wavelength of 750–820 nm. The ICG absorbs a portion of the light within the blood vessels of the target tissue. The absorbed light undergoes a quantum transformation and is released as fluorescent emissions at longer wavelengths, around 820–900 nm. The change from an excited to a ground state occurs within a nanosecond. Emission filters at the camera sensor are used to prevent mixing the unabsorbed excitation light (strong), which is reflected, and the fluorescent light (weak).[29]

Although the fluorescent emission light is only a fraction of the intensity of the excitation light, a surprisingly good signal-to-noise ratio is attained. The image is a brightly fluorescing tissue area composed of blood vessels containing ICG, which can be seen against an almost black background because the filters remove light waves outside the fluorescent range. Without the filters, it would be impossible to see the weak fluorescence image because of the strong reflection of the excitation light 'bouncing back' to the sensor.

The ICG dye is administered intravenously and has a good safety profile. The liver quickly clears it with a half-life of about 3 to 4 minutes. Liver clearance is important as patients with advanced diabetes are prone to renal impairment, which is frequently a contraindication for those agents dependent on renal clearance. One in 42,000 cases of ICG use may experience minor side effects such as hot flashes, hypotension, tachycardia, dyspnea, and urticaria. The frequencies of mild, moderate, and severe side effects were only 0.15%, 0.2%, and 0.05%, respectively. For the competitor substance fluorescein, the proportion of people with side effects is 4.8%.[30]

ICG has several useful properties for studying microcirculation:

- It has been thoroughly verified during its prolonged clinical use;
- It binds efficiently to blood lipoproteins and stays in the vascular space; it does not leak from circulation like fluorescein, which makes it ideal for angiogram;
- It has excellent safety profile – nontoxic and nonionizing, with 60 years of clinical experience;
- It has short half-life, which permits repeated examinations if needed;
- It has excellent signal-to-noise ratio because there is little tissue autofluorescence from the excitation light; and
- It allows deep imaging – to a depth of 2 cm.

ICG angiography is frequently used to monitor flaps in compromised diabetic patients requiring foot or leg amputations. Such flaps are used to close amputation sites. In recent years, ICG has shown promise as a means of monitoring healing. Microangiogenesis is cited as a beneficial mechanism of HBO therapy. A prospective study found ICG angiography to be a useful biomarker for the early identification of responders after their first two HBO sessions. The authors found 100% of the wounds that demonstrated improved perfusion from session 1 to session 2 went on to heal within 30 days of HBO therapy completion, compared with none in the subgroup that did not demonstrate improved perfusion (p <0.01). This study shows the beneficial impact of HBO therapy on perfusion in chronic wounds by ameliorating hypoxia, improving angiogenesis, and the potential role of ICG angiography in identifying those who would benefit from HBO therapy.[31]

3.4 MICROBIOLOGICAL STUDIES

All chronic wounds, and diabetic wounds particularly, are prone to bacterial infections. Wound healing always occurs in the presence of bacteria, whose role depends on their concentration, species composition, and host response.

There are several distinct levels of bacteria present in the wound: contamination (the presence of nonreplicating organisms), colonization (replicating microorganisms adherent to the wound in the absence of injury to the host), and infection. Contamination and colonization by low concentrations of microbes are believed to be normal and do not inhibit wound healing. However, critical colonization and infection are considered to be associated with a significant delay in wound healing.

The switching point from colonization to local infection is a matter of ongoing debate. However, most studies suggest that this transition happens in the range of 10^4–10^5 CFU/g;[32] here, CFU stands for Colony-Forming Units. For example, Breidenbach et al.[33] demonstrated that a bacterial load must exceed 10^4 CFU/g to cause infection in complicated lower-limb wounds. Thus, quantifying the bacterial presence (at least semiquantitative or categorical) is of great importance for a proper diagnosis and treatment selection.

The microbial species composition in wounds changes over time. Quite often, infections are polymicrobial. Group *B Streptococcus* and *Staphylococcus aureus* are common in diabetic foot ulcers.

Early infection diagnostics represent a significant clinical challenge. The current diagnostic approach involves a bedside visual assessment to detect the Clinical Signs and Symptoms (CSS) of wound infection, which may often be supported by semiquantitative microbiological analysis.[34] Currently, the gold standard specimen collection method is to perform a tissue biopsy or needle aspirate of the wound's leading edge after debridement. However, the practical standard for sample collection is a microbiological swab (Levine or Z technique). After specimen collection, the sample is analyzed using culture methods (e.g. pour or spread plate). This method has several significant drawbacks including,(a) a sample can be contaminated by normal skin or mucosa flora, (b) swabs frequently yield too small a specimen for accurate microbiologic examination,[35] and (c) the duration of incubation of cultures can be relatively long. While most aerobic bacteria will grow over two days, anaerobes take weeks and frequently may not grow at all.

NOTES

1. JW Albers, R Pop-Busui, "Diabetic neuropathy: Mechanisms, emerging treatments, and subtypes", *Curr Neurol Neurosci Rep*, 14:473 (2014).
2. R Pop-Busui, AJM Boulton, EL Feldman, VV Bril, "Diabetic neuropathy: A position statement by the American diabetes association", *Diabetes Care*, 40(1):136–154 (2017).
3. EJ Boyko, JH Ahroni, V Stensel, et al. "A prospective study of risk factors for diabetic foot ulcer. The Seattle Diabetic Foot Study", *Diabetes Care*, 22(7):1036–1042 (1999).
4. SJ Rith-Najarian, T Stolusky, DM Gohdes, "Identifying diabetic patients at high risk for lower-extremity amputation in a primary health care setting. A prospective evaluation of simple screening criteria", *Diabetes Care*, 15(10):1386–1389 (1992).
5. H Pham, DG Armstrong, C Harvey, et al. "Screening techniques to identify people at high risk for diabetic foot ulceration: A prospective multicenter trial", *Diabetes Care*, 23(5):606–611 (2000).
6. R Yong, TJ Karas, KD Smith, O Petrov, "The durability of the Semmes-Weinstein 5.07 monofilament", *J Foot Ankle Surg*, 39(1):34–8 (2000).
7. AT Hirsch, MH Criqui, D Treat-Jacobson, et al. "Peripheral arterial disease detection, awareness, and treatment in primary care", *JAMA*, 286:1317–1324 (2001).
8. EF Bernstein, A Fronek, "Current status of non-invasive tests in the diagnosis of peripheral arterial disease", *Surg Clin North Am*, 62:473–487 (1982).
9. X Guo, J Li, W Pang, et al. "Sensitivity and specificity of ankle-brachial index for detecting angiographic stenosis of peripheral arteries", *Circ J*, 72:605–610 (2008).
10. N Diehm, S Rohrer, I Baumgartner, et al. "Distribution pattern of infrageniculate arterial obstructions in patients with diabetes mellitus and renal insufficiency implications for revascularization", *Vasa*, 37:265–273 (2008).
11. L Graziani, A Silvestro, V Bertone, et al. "Vascular involvement in diabetic subjects with ischemic foot ulcer: A new morphologic categorization of disease severity", *Eur J Vasc Endovasc Surg*, 33:453–460 (2007).
12. CV Feen, FS Neijens, SD Kanters, et al. "Angiographic distribution of lower extremity atherosclerosis in patients with and without diabetes", *Diabet Med*, 19:366–370 (2002).
13. SP Nicolai, LM Kruidenier, EV Rouwet, et al. "Ankle brachial index measurement in primary care: Are we doing it right?", *Br J Gen Pract*, 59:422–427 (2009).

14. Y Igarashi, T Chikamori, H Tomiyama, et al. "Clinical significance of inter-arm pressure difference and ankle-brachial pressure index in patients with suspected coronary artery disease", *J Cardiol*, 50:281-289 (2007).
15. D Aerden, D Massaad, K von Kemp, et al. "The ankle–brachial index and the diabetic foot: A troublesome marriage", *Ann Vasc Surg*, 25(6):770-777 (2011).
16. JW Severinghaus, PB Astrup, "History of blood gas analysis. IV. Leland Clark's oxygen electrode", *J Clin Monit*, 2(2): 125-139(1986).
17. JL Mills Sr, MS Conte, DG Armstrong, et al. "The society for vascular surgery lower extremity threatened limb classification system: Risk stratification based on wound, ischemia, and foot infection (WIfI)", *J Vasc Surg*, 59(1):220-234 (2014).
18. CE Fife, C Buyukcakir, GH Otto, et al. "The predictive value of transcutaneous oxygen tension measurement in diabetic lower extremity ulcers treated with hyperbaric oxygen therapy; a retrospective analysis of 1144 patients", *Wound Rep Regen*, 10:198-207 (2002).
19. CE Fife, DR Smart, PJ Sheffield, et al. "Trans transcutaneous oximetry in clinical practice: Consensus statements from an expert panel based on evidence", *UHM*, 36(1):43e53 (2009).
20. DR Smart, MH Bennett, SJ Mitchell, "Transcutaneous oximetry, problem wounds and hyperbaric oxygen therapy", *Diving Hyperb Med*, 36:72-86 (2006).
21. KA Arsenault, A Al-Otaibi, PJ Devereaux, et al. "The use of transcutaneous oximetry to predict healing complications of lower limb amputations: A systematic review and meta-analysis", *Eur J Vasc Endovasc Surg*, 43:329-336 (2012).
22. O Kawarada, Y Yokoi, A Higashimori, et al. "Assessment of macro and microcirculation in contemporary critical limb ischemia", *Catheter Cardiovasc Interv*, 78:1051-1058 (2011).
23. G Urabe, K Yamamoto, A Onozuka, et al. "Skin perfusion pressure is a useful tool for evaluating outcome of ischemic foot ulcers with conservative therapy", *Ann Vasc Dis*, 2:21-26 (2009).
24. MV Marshall, JC Rasmussen, IC Tan, et al. "Near-infrared fluorescence imaging in humans with indocyanine green: A review and update", *Open Surg Oncol J*, 2:12-25 (2010).
25. BD Lepow, D Perry, DG Armstrong, "The use of SPY intra-operative vascular angiography as a predictor of wound healing", *Podiatry Manag*, 30:141-148 (2011).
26. K Igari, T Kudo, T Toyofuku, et al. "Quantitative evaluation of the outcomes of revascularization procedures for peripheral arterial disease using indocyanine green angiography", *Eur J Vasc Endovasc Surg*, 46:460-465 (2013).
27. T Lo, R Sample, P Moore, P Gold, "Prediction of wound healing outcome using skin perfusion pressure and transcutaneous oximetry: A single center experience in 100 patients", *Wounds*, 21:310-316 (2009).

28. T Yamada, T Ohta, H Ishibashi, et al. "Clinical reliability and utility of skin perfusion pressure measurement in ischemic limbs – comparison with other noninvasive diagnostic methods", *J Vasc Surg*, 47:318–323 (2008).
29. B Yuan, NG Chen, Q Zhu, "Emission and absorption properties of indocyanine green in intralipid solution," *J Biomed Opt*, 9(3): 497–503 (2004).
30. M Hope-Ross, LA Yannuzzi, ES Gragoudas, et al. "Adverse reactions due to indocyanine green", *Ophthalmology*, 101(3):529–533 (1994).
31. B Babak Hajhosseini, GJ Chiou, SS Virk, et al. "Hyperbaric oxygen therapy in management of diabetic foot ulcers: Indocyanine green angiography may be used as a biomarker to analyze perfusion and predict response to treatment", *Plast Reconstr Surg*, 147(1): 209–214 (2021).
32. PG Bowler, BI Duerden, DG Armstrong, "Wound microbiology and associated approaches to wound management", *Clin Microbiol Rev*, 14: 244–269 (2001).
33. WC Breidenbach, S Trager, "Quantitative culture technique and infection in complex wounds of the extremities closed with free flaps", *Plast Reconstr Surg*, 95: 860–865 (1995).
34. N Farhan, S Jeffery, "Diagnosing burn wounds infection: The practice gap & advances with moleculight bacterial imaging", *Diagnostics*, 11: 268 (2021).
35. JA Washington, "Principles of diagnosis", In S Baron (editor), *Medical Microbiology*, U of Texas Medical Branch at Galveston; Galveston (TX), 1996, 4th edition.

CHAPTER 4

Anatomic Imaging

As mentioned in the Introduction, by anatomical imaging, we refer to the broad range of techniques that are based on the acquisition of images that are used in wound monitoring or surveillance with the purpose of the systematic collection, analysis, and dissemination of accurate data about wound behavior to improve healing outcomes. Anatomical imaging is based on traditional RGB or monochrome imaging integrated into the clinical workflow. In a typical anatomical imaging scenario, a healthcare provider takes an image of the wound using a Digital Single-Lens Reflex (DSLR) camera or a smartphone. This image is then either stored in an Electronic Health Record (EHR) or processed, and some geometrical information (e.g., wound area, length, width, depth) is retrieved automatically, semi-automatically, or manually. Then, this information can be used to track the healing progress in time. Thus, anatomical imaging is particularly helpful for measuring and documenting wounds.

4.1 WOUND SIZE MEASUREMENT

An important aspect of wound monitoring is documenting, recording, and tracking wound healing progress. Measuring wound size is essential in monitoring the wound-healing process and evaluating treatment effects. It also provides significant value for healthcare documentation, compliance, regulatory issues, and legal protection. For wound measurement, geometrical wound measurements (area and volume) are essential in the wound care armamentarium, as they are objective and can be helpful in cost–benefit analysis.

The focus in the last few decades has been on two-dimensional (2-D) methods to measure wound area. These methods assume that wounds happen in the X–Y plane and that they are flat. However, this assumption is clearly not accurate. For that reason, more recently, three-dimensional (3-D) methods for measuring wound volume have made it possible to evaluate the healing process across all dimensions, including depth which reflects the formation of granulation tissue. Wound size measurement methods are summarized in Table 4.1.

4.1.1 Wound Area Measurements

The wound area measurement methods can be divided into contact methods (e.g., manual and digital planimetry) and non-contact methods (e.g., simple ruler method, mathematical models such as the elliptical method, stereophotogrammetry (SPG) and digital imaging).

4.1.1.1 Planimetry

Planimetric techniques can be manual or digital. In the manual method, a transparent sterile sheet or film[1] is placed on top of the wound, and the margin of the wound is traced with a pen. The tracing is subsequently placed on a metric grid, and the wound area is determined by counting the number of squares covered by the traced area. This method can produce accurate wound measurements; however, it is a very labor-intensive process.

TABLE 4.1 Overview of the wound measurement methods. Reproduced from[a] with modifications

2-D measurement methods	Contact	Planimetric Method (manual and digital)
	Noncontact	Simple ruler method
		Mathematical models
		Stereophotogrammetry
		Digital imaging
3-D measurement methods	Contact	Volume assessment based on depth measurements
		Saline/gel injection into the wound cavity
		Alginate casts
		Kundin device
	Noncontact	Structured light technique
		Stereophotogrammetry
		Laser scanners
		Digital imaging

[a] LB Jørgensen, JA Sørensen, GB Jemec, KB Yderstræde, "Methods to assess area and volume of wounds – a systematic review", *Int Wound J*, 13(4):540–553 (2016).

In early digital planimetric applications (e.g., Visitrak [Smith & Nephew, Hull, UK]), the margin of the wound on a sterile sheet is retraced onto a tablet computer that performs the same calculations. Currently, this method is integrated into digital imaging and has become noncontact. In the authors' opinion, this method should be considered the gold standard to which all subsequent techniques are compared to assess their accuracy.

4.1.1.2 Simple Ruler Method

More commonly, geometrical wound measurements are performed manually using a ruler. In particular, there are two primary methods used for wound measurements:[2]

- *Greatest length and width method: In this method, the greatest length and the greatest width of the wound are measured across the wound's diameter, from wound edge to the opposite wound edge.*

- *Clock method: In this method, the face of a clock is used to guide measurement. The 12:00 reference position is towards the head of the body, and measurements are obtained from 12:00 to 6:00 (length) and from 9:00 to 3:00 (width).*

Notice that the width can be larger than the length in this case.

However, only the length, L, and width, W, can be determined using these methods. The surface area of the wound, S in this case, can be estimated as $S = L \times W$, which is a very rough approximation that does not consider the wound's shape. Because these measurements assume that the wound is a rectangle, this type of calculation has been shown to overestimate the wound area by 10–44%,[3] with accuracy decreasing as wound size increases.[4] Thus, manually tracking the progress of wound healing in time will be relatively imprecise.

4.1.1.3 Mathematical Models

The mathematical model method assumes that most wounds are circular or elliptical in nature, and the area can therefore be calculated using standard mathematical formulae. The most common approach is the elliptical method, in which the area is calculated by multiplying $\pi/4$ by the shortest and longest diameters of the wound.

Mayrovitz et al.[5] investigated shape and area measurements in 81 diabetic plantar ulcers and derived a new area formula $S = 0.73 \times L \times W$

(L=length, W=width). This formula was found to be more accurate than the elliptical model when compared with digital planimetry; however, it is inaccurate in wounds that do not follow an elliptical pattern.

4.1.1.4 Stereophotogrammetry

This method uses a stereographical camera linked to a computer to capture an image of the wound. Once the image is downloaded to the computer, the wound perimeter is traced by moving the cursor on the monitor. Then, the computer software calculates the wound area, length, and width. Such systems include SPG (Vista Medical, Winnipeg, Manitoba, Canada) or LifeViz (QuantifiCare, Sophia Antipolis, France).

4.1.1.5 Digital Imaging

The accuracy of wound measurements can be increased with digital photography. In this case, the precise wound area can be calculated in addition to the more accurate length and width measurements (planimetry).

The wound can be traced manually, semi-automatically, or automatically. Semiautomatic segmentation involves software automatically segmenting the wound and then having a healthcare professional verify and correct the results. This method is faster than manual segmentation and is less prone to inter-observer variability, but it is still time-consuming and can be affected by image quality. The automated method involves unsupervised auto-segmentation. This method is faster than other methods and is less prone to inter-observer variability, but can be affected by image quality.

In particular, digital technology may lead to a 10× increase in the accuracy of wound measurements. However, the initial implementation of such techniques using DSLR cameras received limited clinical traction, primarily due to the significant extra time required to take pictures using specialized equipment. Thus, this process did not fit well in a busy clinical workflow. With the advent of smartphones, wound management was revolutionized. The ability to capture an image and annotate the wound using the same tool significantly improved the overall wound management workflow.

Multiple studies compared different wound area measurement methods. While many studies demonstrated close results for smaller wounds (less than 10cm^2), the discrepancy appears for larger wounds. For example, Oien et al.[6] compared four methods of wound area measurement (digital planimetry, mechanical planimetry, grid tracing (manual planimetry), and simple ruler method) in 20 patients with 50 chronic leg ulcers of various etiology. All the methods demonstrated a high degree of agreement in wounds less than 10 cm^2 ($p<0.01$), but differences occurred with increasing wound size. Nonetheless, another study[7] found that digital wound area methods had an inter-rater reliability of 0.99, indicating that, with the appropriate tools, this method is highly reproducible even in moderately trained operators.

Contact planimetric methods (either manual or digital) can be considered to be ground truth in wound area measurements as they can accommodate different body and wound geometries, including circumferential wounds. However, they require contact with the wound tissue and are very labor-intensive. The digital imaging methods are probably the second best in accuracy. For example, Thawer et al.[8] compared the intra- and inter-rater reliability of digital planimetry and digital imaging methods in measuring chronic wounds ($n=45$). The techniques were equally reliable, based on similar ICC and standard measurement errors.

However, digital imaging is a non-contact method and much superior from the integration in the workflow perspective. For example, a study found a reduction in over 50% of the time required to complete a wound's measurement when using digital tools, compared to traditional methods, indicating a potential time-cost saving in the clinical workflow.[9] Furthermore, with constantly evolving features (like automated wound margin delineation), digital imaging is well-positioned to become the standard of care (Figure 4.1).

4.1.2 Wound Volume Measurements

Wound volume measurements are particularly important as they reflect granulation tissue formation and the closure of the wound from the bottom to the top. Furthermore, these methods take into account the fact that wounds are 3-D structures and not 2-D planes. The wound volume measurement techniques can be split into contact and noncontact ones.

4.1.2.1 Contact Methods

The wound volume can be estimated based on wound depth measurements.

The wound depth can be measured in several ways. Most commonly, for depth measurements, a sterile swab is inserted into the deepest area of the wound. Then, the height parallel to the external boundary of the ulcer is observed and marked to determine the wound depth, D.

Anatomic Imaging ■ 33

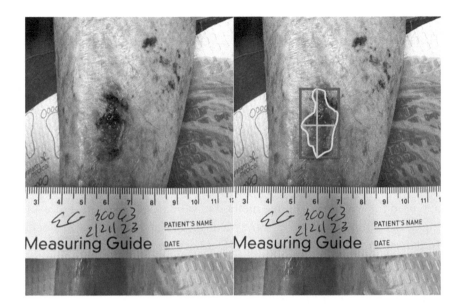

FIGURE 4.1 Wound area measurements. The most common wound area estimation method consists of identifying the wound's major and minor axis (green lines), measuring them, and estimating the area based on this (red box). However, as noted, the real wound's edge (yellow line) is overestimated when using this method. This has particularly been shown to become more problematic as the size of a wound increases. Therefore, digital planimetry with fiducial markers is emerging as the gold standard for wound care.

Another approach to measure wound depth is using a Kundin® gauge (Pacific Technologies and Development Corporation, San Mateo, CA), a ruler-based device with three disposable paper rulers set at orthogonal angles to measure the wound's length, width, and depth.

Once the wound depth has been measured, the wound volume can be estimated using measured wound dimensions L, W, and D. Occasionally, the simple formula $L \times W \times D$ is used. However, typically it grossly overestimates the wound volume, as this assumes a constant depth throughout the wound. Thus, various refinements were proposed. For example, Langemo et al.[10] used the formula $L \times W \times D \times 0.327$ to account for depth variations.

However, manual depth measurements are subject to significant interrater variability. Moreover, multiple studies show that depth-based wound volume assessments underestimate or overestimate the wound volume as wounds do not have a uniform depth. As such, direct measurements of the wound volume are more objective.

There are several direct contact methods to measure wound volume. They include saline gel injection into a wound cavity and making a mold. Saline gel injection into a wound cavity is inexpensive but labor-intensive and carries the risk of wound contamination. Another contact wound volume measurement method is to fill the wound with silicone or alginate to make a mold. Then, the material is placed in a beaker, where the water displacement indicates wound volume. However, this method is even more labor-intensive and also carries the risk of wound contamination. Still, these methods are used occasionally as a reference to validate non-contact volume measurements, especially in animal models. For example, Davis et al.[11] used the weight-to-volume method to validate a non-contact measuring device Silhouette Star® (Aranz Medical, Christchurch, New Zealand). In particular, they found that this 3-D device underestimated depth and volume ($p<0.05$).

4.1.2.2 Non-contact Methods

Non-contact optical methods for wound volume measurements include the laser triangulation method (laser scanner), structured light, and stereophotogrammetry.

The laser scanner working principle is based on optical triangulation, obtained by projecting a collimated laser beam onto a target and acquiring the profile shape with an imaging device that must be placed at a specific, known angle. The scanning is typically achieved by projecting a laser beam on a rotating mirror (a polytope). Laser scanners have multiple applications in 3-D profilometry. Thus, the commercially available 3-D profilometers can be repurposed for wound profile measurements. However, this modality is used for clinical research and has not yet been translated into routine clinical practice. For example, Romanelli et al.[12] used a Vivid 900 laser scanner (Minolta, Osaka, Japan) to scan and digitize the wound shape. In a study on 15 venous leg ulcers, the authors found excellent intra-rater and inter-rater reproducibility with a very low relative error value. The mean ± SD time for a full scan acquisition on the wound area and volume was 3.6 ± 1.4 minutes.

Structured light or active illumination profilometry techniques illuminate the measurement object with predefined spatially varying intensity light patterns (typically lines). Then, distortions from the straight lines can be used to reconstruct the surface shape. Structured light techniques have multiple applications in 3-D profilometry. For example, in wound care, in addition to several experimental modalities, there are commercially available medical devices like SilhouetteStar (Aranz Medical, Christchurch, New Zealand), which uses structured lighting for 3-D measurements of the wound area, depth, and volume.

Stereophotogrammetric (SPG) systems simultaneously capture two digital wound images at different angles. Then a software application creates a 3-D reconstruction of the wound. SPG systems have been translated into medical practice. For example, LifeViz® (QuantifiCare, San Mateo, CA) product line uses SPG for aesthetic surgery applications. For face reconstruction, it takes three photos, which are then automatically stitched, and the patient's face is reconstructed in 3-D. This system was tried in wound care by Davis et al.[13] who found high inter-rater reliability for volume measurements (ICC=0·9867; $p<0.001$). Furthermore, compared with the simple ruler method, only width measurements showed a significant difference ($p<0.0001$), while surface area, depth, and length values were similar. Recently, stereophotogrammetric imaging was implemented into the smartphone app Swift Skin & Wound (Swift Medical, Toronto, Canada), which does not require specialized optics and hardware.

With recent advances in Time-of-Flight (ToF) cameras, one could expect that they will also be used for depth and volume wound measurements.

Despite its promise, wound volume measurements still face numerous challenges. One of the factors affecting the accuracy of wound measurement is the definition of the wound boundary, which is often challenging to identify. For example, localization on a curved part of the body (e.g., the heel) can make it particularly difficult to estimate wound size correctly. In this case, some anatomical models need to be used to predict the tissue boundary in the absence of the wound. Then, the wound volume can be found as the volume of the missing part.

In addition, several other factors complicate accurate wound volume measurements, including that some wounds are also extensively undermined, making the volume challenging to assess; the presence of necrotic tissue, which can be difficult to measure accurately, thus leading to errors in volume calculations; and wounds in areas with thick soft tissue that can pose problems because of physiological contraction or fibrotic scar formation.

4.2 TISSUE COMPOSITION

Tissue composition is another critical aspect of wound surveillance and progression monitoring. Wounds are composed of four main tissue types: epithelial tissue, granulation tissue, slough, and eschar, which are described in more detail in Appendix A.

The knowledge of the wound's tissue composition is quite important. For example, the devitalized tissue often prevents wound healing and needs to be removed. Typically, the composition of the tissue is estimated visually (in percent) by a healthcare practitioner. However, it has been well established that visual estimation is very inaccurate, particularly for wounds with complex shapes. For example, a study[14] showed very low inter-rater agreements for the tissue type area estimation, despite these clinicians scoring very high intra-rater agreement scores. Thus, automated methods can be a great help for more objective measurements.

In a typical scenario of measurement automatization, the information is extracted from the digital image, and the tissues present are classified. To some extent, the different wound tissues have 'absolute' colors. By 'absolute,' we mean their color does not depend on their skin tone. This makes the tissue classification very different from other tissue classification tasks. For example, erythema demonstrates 'relative' colors, as its appearance depends strongly on skin tone.

Given these 'absolute' colors, both deep learning and simple coloristic approaches are suitable for tissue segmentation and classification. The typical approach for tissue segmentation and classification is to use Machine Learning techniques. For example, the CNN-based auto-classification feature, AutoTissue, has been implemented in Swift Skin & Wound (Swift Medical, Toronto, Canada). Using the four types of wound tissues described above, this model demonstrated a mean Intersection Over Union (IOU) of 0.8644 and 0.7192 for wound and tissue segmentation. Additionally, its predictions were rated by a consensus of trained physicians to be accurate in 91% of the cases.[14]

4.3 WOUND HEALING TRAJECTORY ASSESSMENT

Wound closure is the most objective measure of wound healing. In particular, wound closure is the single parameter used by the FDA to measure wound therapeutics' efficiency. However, wound closure for a chronic wound can be lengthy. Thus, surrogate wound healing measures, which evaluate the effectiveness of the intervention at a particular time point, are required.

The wound area, S, is an important clinical indicator of the wound status and can be used to predict wound healing progress and clinical outcomes. In particular, S is a part of several wound indices (e.g., PUSH score for pressure injuries). Change in wound area is the most common metric to track the wound healing progress. While many variations of this measure exist (see[15] for details), most commonly, a measure of the change in wound area is used and is expressed either as a raw number (cm^2) or as a percentage of the initial wound area.

However, it is known that the wound healing rate expressed as the absolute area healed per day tends to exaggerate the healing rates of larger wounds. Similarly, the healing rate expressed as a percentage of the initial area healed per day tends to exaggerate smaller wounds' healing rates.[15]. Thus, more objective methods need to be adopted for clinical use.

One such metric could be a linear healing rate D (termed initially as 'the advance of the wound margin toward the wound centre') proposed in,[16] which can be calculated as $D=\Delta S/P$ from the change in the area ΔS and mean perimeter P. Gorin et al.[17] in a study on 49 wounds, found that the linear healing rate does not correlate with the initial wound area, perimeter, and W/L ratio. Thus, the linear healing rate is independent of the wound shape. These results were confirmed later by[15] on 300 wounds.

The linear healing rate was further assessed in other studies. Pecoraro et al.[18] found 0.064 mm/day in diabetic foot patients. Margolis et al.[19] found 0.093 mm/day on venous ulcers. Gorin et al.[17] found a similar 0.11 mm/day result on venous ulcers. Finally, Cukjati et al.[15] found 0.068 mm/day for wounds of unknown etiology. All these values are two to four times lower than the angiogenesis-limited healing rate. Thus, these rates reflect limited or slower collagen deposition processes or areas with impaired healing.

It has been known that the wound shape also affects the healing trajectory. For example, in a study on 338 Venous Leg Ulcers (VLU), Cardinal et al.[20] found that VLUs that transitioned to a more convex wound shape, and maintained a linear relationship between their wound margin size and wound surface area size, had faster healing rates and were more likely to completely heal by 12 weeks (odds ratio=4.84, p=0.001).

Recently, approaches that consider wound shape have been explored. For example, Vidal et al.[21] proposed to use the continuous linear healing rate (D_c) instead of the linear healing rate (D). Using simple calculations, they obtained a regression model $S/P= D_c t +q$. Here, q depends on the wound geometry. This model has been extended to more complex shapes.[22]

Bull et al.[23] found that the healing of 40 Venous Leg Ulcers receiving multicomponent compression bandaging followed a linear trajectory over four weeks, as measured by gross area healed, percentage area healed, and advance of the wound margin.

Gupta et al.[24] proposed a fully automatic model for predicting wound closure based on objective AI-based measurements. This model incorporated the wound's area, its change in size from the previous measurement, tissue composition, and wound edge characteristics and was developed from a set of 2.1 million wound measurements. Compared to PUSH and BWAT scores, its clinical performance was found to be significantly more robust and to create a predicted healing trajectory, thus enabling clinicians to determine whether a particular wound is following its expected trajectory. However, despite its initial success, more research on the clinical application of such models is still needed for widespread adoption.

As a final note, from a wound healing trajectory assessment perspective, it is vital to have wound measurements at multiple time points (continuous measurements). In addition to providing baseline information, continuous measurements help to predict healing and aid in monitoring treatment efficacy and evaluation.

4.4 WORKFLOW EFFICIENCY

Workflow efficiency in wound care refers to optimizing how wound care is provided to patients, aiming to maximize the quality of care provided while minimizing the time and resources required to provide it. Efficient wound care workflows typically involve a structured approach to patient assessment, wound diagnosis, and treatment planning, which helps to ensure that the proper treatment is delivered at the right time.

Efficient wound care workflows can help reduce the time and resources required to provide care while improving patient outcomes. By streamlining the wound care process, healthcare providers can ensure that patients receive timely and effective treatment, which can help to improve healing rates, reduce complications, and enhance the overall quality of care.

Thus, the primary way to achieve these goals is workflow automation. Workflow automation allows streamlining processes, sharing best practices, and saving time. It also allows for complete data visibility and can help healthcare organizations meet compliance and regulatory regulations.

Optical imaging technologies are instrumental in achieving these automation goals as they can be easily integrated into the clinical workflow and support workflow automation. One particular area of efficient workflow is documentation. Healthcare professionals are often overburdened with various documentation, which needs to be entered into multiple systems.

Thus, developing an efficient workflow, which saves time and reduces possible user errors, is of great importance.

Early imaging systems for wound documentation included DSLR cameras. However, they had a significant downside as images needed to be manually transferred from the camera to the computer and uploaded into the EHR. Moreover, there are substantial hurdles regarding patient data privacy when using this approach. With the advances in technologies and the proliferation of smartphones, current wound imaging systems are integrated with EHR systems already and allow capturing all additional wound information alongside wound imaging in secure digital environments.

Several studies have shown the impact of optical imaging technologies on workflow automation and streamlining. For example, Wang et al.[25] found that over 80% of patients in a cohort reported that photographing their wounds helped them track their progress and led to them being more involved in their care in over 50% of the cases. The same author group later found that digital wound imaging reduced approximately 60% of the time needed to measure a wound, with further time savings expected for complete wound documentation if this system is linked to an EHR.[26]

A time–motion study[27] measured the total time required for wound assessments in a real-world environment using a digital application vs. manual methods. The study also compared the proportion of images with acceptable quality on the first attempt for the two methods.

Assessment activities included: labeling wounds, capturing images, measuring wounds, calculating surface areas, and transferring data into the patient's record. A total of 115 wounds were assessed. The average time required to capture and access wound images with the digital AI tool was significantly faster than a standard digital camera, with an average of 62 seconds ($p<0.001$). The digital application was markedly faster by 77% at accurately measuring and calculating the wound surface area with an average of 45.05 seconds ($p<0.001$). Overall, the average time to complete an AI-powered digital wound assessment was significantly faster by 79%.

Using the AI application, the staff completed all the steps in about half the time (54%) normally spent on manual wound evaluation activities. Moreover, acquiring an acceptable wound image from the first attempt was significantly more likely to be achieved using the digital tool than the manual methods (92.2% vs. 75.7%, $p<0.004$).[27]

Another research group demonstrated that optical imaging allows efficient treatment selection in burn victims and that digital measurements can be used to select the treatments these patients are most likely to benefit from.[28] Finally, research shows a reduction from 12.9% to 2.9% in the incidence of facility-acquired pressure injuries in a skilled nursing home after adopting digital imaging technologies as part of a quality improvement program.[29] Together, these results greatly highlight how workflow automation significantly affects the quality of care.

4.5 TELEMEDICINE

Telehealth or Telemedicine (TM) is an ambiguous literature term. It encompasses a variety of healthcare interventions such as video conferencing, digital stethoscopes, text messages, behavioral apps, and real-time physiologic monitoring with alerts for changes in weight, temperature, tissue oxygen, and other physiologic parameters.[30] However, it may have significant implications in wound care.

Streaming video consultations and image sharing with patients has been studied and may reduce hospitalization and amputation rates. Two meta-analyses of studies using specialist nurses or wound care experts reported that TM management improved the healing of a heterogeneous group of wounds, including diabetes ulcers, compared to standard care (Hazard Ratio, HR = 1.40, $p = 0.01$).[31] There was a significant reduction in the frequency of amputation after 12 months in the TM group compared to the standard care groups (Risk Ratio, RR = 0.52, $p = 0.003$). Notably, meta-analyses reported no significant difference in mortality between telehealth and standard care.[32] Other reviews reported faster ulcer healing rates using a telephone-based advisory system or home nurses to assist in telemonitoring and non-inferiority for time to healing and amputation rate.[33]

Three of the four randomized controlled trials reported that daily remote physiologic monitoring (RPM) of infrared foot temperature monitoring of hot spots significantly reduced the risk of recurrent diabetes-associated foot ulcers compared to standard care. Thermometric mats had high sensitivity in detecting hot spots that appear before a diabetic ulcer. With early detection comes early interventions, which effectively prevent many diabetic ulcers.[34]

Kong et al.[35] described an innovative remote care strategy consisting of a patient-facing wound care Swift Medical Patient Connect App (Swift Medical, Toronto, Canada) designed to monitor and manage wounds by a patient with diabetes and foot ulcers. They reported that patients found the technology 'educational and empowering,' with increased self-examination and engagement in preventive behavior such as monitoring for trauma and early signs of infection.

Technical challenges, such as slow or unstable internet, poor lighting, small screens, and poor sound quality, affected whether the patients experienced the TM solution as beneficial for their wound treatment. When implementing virtual imaging solutions, attention was also drawn to the need to consider human factors such as retinopathy and/or neuropathy.[36]

Modern Telemedicine represents a quantum leap from the early days of virtual care delivery. Initial Telemedicine solutions offered simple two-way video sessions where a clinician would diagnose patients based on their symptoms. However, wound care is a 'visually demanding' specialty, and video alone cannot meet the clinical requirements. Professionally calibrated still images of the wound are essential. Video fails to provide 'true color' corrections; video cannot be dynamically enlarged to inspect the wound bed; the compressed nature of video creates jagged, pixelated images when enlarged; and video lacks the basic requirements of 'store and forward' which is essential for cross consultation, communication, and coordination (see Figure 4.2).

Finally, video cannot provide a standardized, discrete comparison of the wound bed over time. To do so requires the images to be created with uniform distance, rotation, color correction, area measurements, and time

FIGURE 4.2 Video lacks the capacity for 'macro' viewing of an enlarged image for inspecting detailed anatomy within the wound bed. Because video is highly compressed, any enlargement leads to immediate 'pixelation' and loss of clinically important details.

stamps – none of which can be done with video. The video channel is useful for communication, and the 'still image' channel is essential for analysis and quantitative comparisons over time.

In summary, the field of image-based virtual care is rapidly evolving. Meta-analysis of studies using video sessions or uncalibrated images shows favorable benefits or at least 'non-inferiority' outcomes. Future studies using 'enhanced video sessions' combining video with store-and-forward, color-corrected, calibrated images (see section 4.6) will demonstrate further benefits. Similarly, Remote Patient Monitoring (RPM) using infrared imaging will provide early warning of impending ulceration.

4.6 UNDERSTANDING COLOR BIAS

Understanding color bias is fundamental to a successful evaluation, particularly in Telemedicine. Because lighting conditions in the home are nonuniform, the physician must have the means to compensate for color bias digitally. Failure to correct color bias will lead to incorrect clinical assessments. For example, a red bias could falsely suggest infection in some cases. A green bias could mean certain types of bacterial infection, e.g., *Pseudomonas*. A blue bias could mean poor blood flow or bruising (see Figure 4.3).

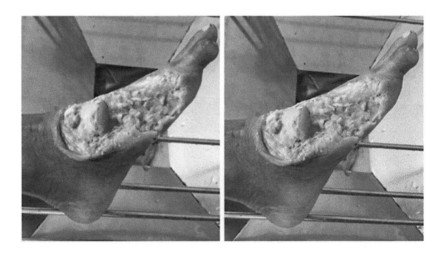

FIGURE 4.3 The image on the left is color corrected. Note the presence of red tissue and pink toes in the corrected image and the yellow/green hues of the tissue and toes in the right image. Only the color-corrected image on the left could be used for clinical decision-making.

The human perception of net 'color' is a function of color temperature and illumination. The Kelvin scale ranks the color temperature or 'warmth' of light. The 'warmer' the color, the lower the Kelvin temperature. For example, tungsten light bulbs have a red/yellow cast and are in the range of 2,000–3,000 K. Standard fluorescent lights have a blue/green cast and are in the range of 6,000–8,000 K. However, the perception of color is also a function of illumination measured in lux.

When choosing or correcting for a desired net color to achieve a natural, clinically correct color, it is important to stay within the boundaries of the empirically determined Kruithof diagram (Figure 4.4). Using high color temperatures at low illumination levels results in colors appearing flat and dull. Conversely, the use of warm color temperatures at high illumination levels yields an unpleasant yellow hue to the image, disturbing the feeling of well-being. The matching of the illumination level to the desired color temperature is a critical point that is essential to professional-grade imaging for diagnostic purposes.

The Kruithof envelope was constructed from psychophysical data collected by Dutch physicist Arie Andries Kruithof (see Figure 4.4). Lighting

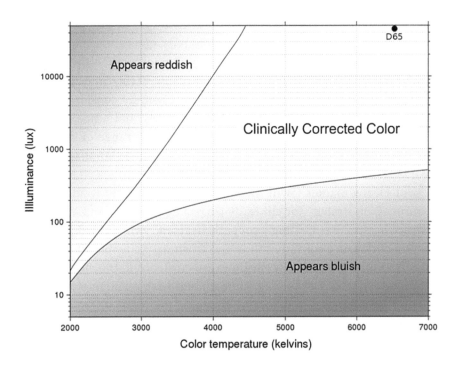

FIGURE 4.4   Kruithof curve. Modified from [37] under CC BY-SA 4.0 license.

conditions within the bounded region were empirically assessed as natural, whereas conditions outside the area were considered displeasing or unnatural.[37]

For example, natural daylight has a color temperature of 6,500 K and an illuminance of about 10^4 to 10^5 lux. This color temperature–illuminance pair results in a natural color rendition, but it would appear bluish if viewed at low illuminance. At typical indoor office illuminance levels of about 400 lux, pleasing color temperatures are lower (between 3,000 K and 6,000 K), and at typical home illuminance levels of about 75 lux, pleasing color temperatures are even lower (between 2,400 K and 2,700 K). These color temperature–illuminance pairs are often achieved with fluorescent and incandescent sources. The 'natural' region of the curve contains color temperatures and illuminance levels comparable to color-balanced, natural light.

Kruithof's findings are directly related to human adaptation to changes in illumination. As illuminance decreases, human sensitivity to blue light increases, known as the Purkinje effect.[38]

When luminance levels decrease, the human visual system switches from photopic (cone-dominated) vision to scotopic (rod-dominated) vision. Rods have a very high spectral sensitivity to blue energy, whereas cones have varying spectral sensitivities to reds, greens, and blues. Since the dominating photoreceptor in scotopic vision is most sensitive to blue, human sensitivity to blue light increases. Because of this, intense sources of higher (bluer) color temperatures are all generally considered to be unnatural or 'displeasing' at low luminance levels, and a narrow range of clinically correct sources exist. Subsequently, the range of diagnostically acceptable sources increases in photopic vision as luminance levels are increased.

Swift HealX marker (Swift Medical, Toronto, Canada) is an example of an imaging strategy for clinically correct color, focus, and a uniform distance perspective. Getting a properly 'framed,' well-focused picture at the proper distance consumes time and frustrates patients and providers alike. If a camera is too close, shadows are created, and the moist surface of a wound creates a distorting white glare if flash photography is used. The app provides user guidance by changing the colors of the displayed marker.

In summary, a color correction strategy must be in place to ensure the image quality of diagnostic caliber upon which reliable clinical and therapeutic decisions can be made.

NOTES

1. DK Langemo, H Melland, D Hanson, et al. "Two-dimensional wound measurement: Comparison of 4 techniques", *Adv Wound Care*, 11(7): 337–343 (1998).
2. L Swezey, "Methods and strategies for accurate wound measurement", 2014, https://www.woundsource.com/blog/5-techniques-accurate-wound-measurements.
3. RJ Goldman, R Salcido, "More than one way to measure a wound: An overview of tools and techniques", *Adv Skin Wound Care*, 15(5):236–243 (2002).
4. C Majeske, "Reliability of wound surface area measurements", *Phys Ther*, 72(2):138–141 (1992).
5. HN Mayrovitz, J Smith, C Ingram, "Geometric, shape and area measurement considerations for diabetic neuropathic plantar ulcers", *Ostomy Wound Manage*, 43:58–62, 4–5 (1997).
6. RF Oien, A Hakansson, BU Hansen, M Bjellerup, "Measuring the size of ulcers by planimetry: A useful method in the clinical setting", *J Wound Care*, 11:165–168 (2002).
7. SC Wang, JAE Anderson, R Evans, et al. "Point-of-care wound visioning technology: Reproducibility and accuracy of a wound measurement app", *PLoS One*, 12(8):e0183139 (2017).
8. HA Thawer, PE Houghton, MG Woodbury, et al. "A comparison of computer-assisted and manual wound size measurement", *Ostomy Wound Manage*, 48:46–53 (2002).
9. Y Au, B Beland, JAE Anderson, et al. "Time-saving comparison of wound measurement between the ruler method and the swift skin and wound app", *J Cutan Med Surg*, 23(2):226–228 (2019).
10. DK Langemo, H Melland, B Olson, et al. "Comparison of 2 wound volume measurement methods", *Adv Skin Wound Care*, 14:190–196 (2001).
11. KE Davis, FC Constantine, EC Macaslan, et al. "Validation of a laser-assisted wound measurement device for measuring wound volume", *J Diabetes Sci Technol*, 7:1161–1166 (2013).
12. M Romanelli, V Dini, T Bianchi, P Romanelli, "Wound assessment by 3-dimensional laser scanning", *Arch Dermatol*, 143:1333–1334 (2007).
13. AJ Davis, J Nishimura, J Seton, et al. "Repeatability and clinical utility in stereophotogrammetric measurements of wounds", *J Wound Care*, 22:90–92, 4–7 (2013).
14. D Ramachandram, JL Ramirez-GarciaLuna, RDJ Fraser, et al. "Fully automated wound tissue segmentation using deep learning on mobile devices: Cohort study", *JMIR Mhealth Uhealth*,10(4):e36977 (2022).
15. D Cukjati, S Reberšek, D Miklavčič, "A reliable method of determining wound healing rate", *Med Biol Eng Comput*,39: 263–271 (2001).
16. T Gilman, "Parameter for measurement of wound closure", *Wounds*, 2:95–101 (1990).

17. DR Gorin, PR Cordts, WW LaMorte, et al. "The influence of wound geometry on the measurement of wound healing rates in clinical trials", *J Vasc Surg*, 23(3):524–528 (1996).
18. RE Pecoraro, JH Ahroni, EJ Boyko, et al. "Chronology and determinants of tissue repair in diabetic lower extremity ulcers", *Diabetes*, 40:1305–1313 (1991).
19. DJ Margolis, EA Gross, CR Wood, et al. "Planimetric rate of healing in venous ulcers of the leg treated with pressure bandage and hydrocolloid dressing", *J Am Acad Dermatol*, 28:418–421(1993).
20. M Cardinal, DE Eisenbud, DG Armstrong, "Wound shape geometry measurements correlate to eventual wound healing", *Wound Repair Regen*, 17(2):173–178 (2009).
21. A Vidal, H Mendieta Zerón, I Giacaman, et al. "A simple mathematical model for wound closure evaluation", *J Am Coll Clin Wound Spec*, 7(1-3):40–49 (2016).
22. G Saiko, "The impact of the wound shape on wound healing dynamics: Is it time to revisit wound healing measures?" In Proceedings of the 14th International Joint Conference on Biomedical Engineering Systems and Technologies (BIOSTEC 2021) –Volume 2: BIOIMAGING, pp. 182–187.
23. RH Bull, KL Staines, AJ Collarte, et al. "Measuring progress to healing: A challenge and an opportunity", *Int Wound J*, 19(4): 734–740 (2022).
24. R Gupta, L Goldstone, S Eisen, et al. "Towards an AI-based objective prognostic model for quantifying wound healing", *IEEE J Biomed Health Inform*, PP (2023).
25. SC Wang, JA Anderson, DV Jones, R Evans, "Patient perception of wound photography", *Int Wound J*, 13(3):326–30 (2016).
26. Y Au, B Beland, JAE Anderson, et al. "Time-saving comparison of wound measurement between the ruler method and the swift skin and wound app", *J Cutan Med Surg*, 23(2):226–228 (2019).
27. HT Mohammed, RL Bartlett, D Babb, et al. "A time motion study of manual versus artificial intelligence methods for wound assessment", *PLoS One*, 17(7):e0271742 (2022).
28. MA Martínez-Jiménez, JL Ramirez-GarciaLuna, ES Kolosovas-Machuca, et al. "Development and validation of an algorithm to predict the treatment modality of burn wounds using thermographic scans: Prospective cohort study", *PLoS One*, 13(11):e0206477 (2018).
29. Y Au, M Holbrook, A Skeens, et al. "Improving the quality of pressure ulcer management in a skilled nursing facility", *Int Wound J*, 16(2):550–555 (2019).
30. JJ van Netten, D Clark, PA Lazzarini, et al. "The validity and reliability of remote diabetic foot ulcer assessment using mobile phone images", *Sci Rep*, 7(1):9480 (2017).
31. Z Huang, S Wu, T Yu, A Hu, "Efficacy of telemedicine for patients with chronic wounds: A meta-analysis of randomized controlled trials", *Adv Wound Care*, 10(2):103–112 (2020).

32. A Drovandi, S Wong, L Seng, et al. "Remotely delivered monitoring and management of diabetes-related foot disease: An overview of systematic reviews", *J Diab Sci Techn*,17(1): 59–69 (2023).
33. L Chen, L Cheng, W Gao, et al. "Telemedicine in chronic wound management: Systematic review and meta-analysis", *JMIR Mhealth Uhealth*, 8(6):e15574 (2020).
34. C Hazenberg, WB Aan de Stegge, SG Van Baal, et al. "Telehealth and telemedicine applications for the diabetic foot: A systematic review", *Diabetes Metab Res Rev*,36(3):e3247 (2020).
35. LY Kong, JL Ramirez-Garcia Luna, R Fraser, SC Wang, "A 57-year-old man with type 1 diabetes mellitus and a chronic foot ulcer successfully managed with a remote patient-facing wound care smartphone application", *Am J Case Rep*, 22: e933879 (2021).
36. MM Iversen, J Igland, H Smith-Strom, et al. "Effect of a telemedicine intervention for diabetes-related foot ulcers on health, well-being and quality of life: Secondary outcomes from a cluster randomized controlled trial (DiaFOTo)", *BMC Endocr Disord*, 20(1):8 (2020).
37. Kruithof curve. In Wikipedia, 2023, May 24, https://en.wikipedia.org/wiki/Kruithof_curve.
38. JP Frisby, *Seeing: Illusion, Brain and Mind*, Oxford University Press, Oxford,1980.

CHAPTER 5

Optical Diagnostic Techniques

ADEQUATE MACRO AND MICROCIRCULATION are necessary for healing. Insufficient perfusion impairs angiogenesis, collagen deposition, and epithelialization leading to sustained inflammation. Hypoxia is a reduction in oxygen delivery against cellular demand. In contrast, ischemia is a state in which perfusion is lacking, resulting in hypoxia and a diminished supply of nutrients needed to repair tissues.[1,2,3,4] Because oxygen is the 'rate-limiting' step for healing, the ideal measurement system would provide a noninvasive, direct measurement of tissue oxygen levels in both the wound and the peri-wound tissues.

Thus, most physiological imaging techniques described in this chapter measure perfusion directly (like Laser Doppler Imaging) or indirectly (like hyperspectral imaging). However, perfusion is only one determinant of outcome; wound size, wound depth, wound tissue composition, and the presence of infection also govern the prognosis for healing or amputation. To facilitate communication and research, the Society for Vascular Surgery (SVS) created a new classification system for the threatened lower extremity incorporating these important considerations. This new system is referred to as the SVS Wound, Ischemia, and Foot Infection classification system or 'WIFI classification' for short. The WIFI system has been clinically validated as a predictor for limb salvage and wound healing.[5]

Practical methods to evaluate wound features were described in the previous chapter (Chapter 4). The following sections of this chapter will review novel and practical ways to evaluate wound perfusion and infection. These methods have been used in clinical research but have not yet been widely adopted in clinical practice. However, most of these technologies have the potential to be used in point-of-care settings and thus can be used for DFU screening.

In Figure 5.1, several techniques have been depicted in resolution-sampling depth coordinates.

Physiological imaging methods in wound care can be roughly split into two categories: diagnostic technologies that target abnormalities of blood circulation (perfusion) and technologies that aim to detect infection.

5.1 PERFUSION TARGETING TECHNIQUES

Perfusion refers to the fluid movement through the circulatory or lymphatic system toward a particular organ or tissue, typically in reference to the blood supply to capillary beds within tissues. The measurement

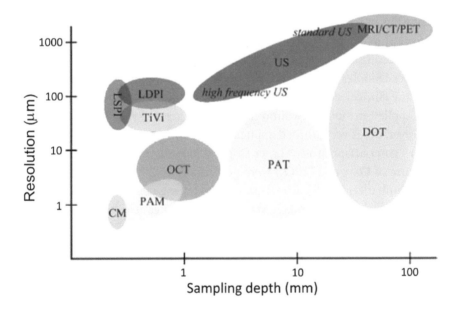

FIGURE 5.1 The relative domains occupied by various imaging modalities in terms of resolution and sampling depth. DOT – Diffuse Optical Tomography, US – ultrasound, LDPI – Laser Doppler Perfusion Imaging, PAT – Photoacoustic Tomography, LSPI – Laser Speckle Perfusion Imaging, OCT – Optical Coherence Tomography. (Reproduced from with permissions.) SM Daly, MJ Leahy, 'Go with the flow': A review of methods and advancements in blood flow imaging", *J Biophotonics*, 6(3): 217–255 (2013).

of perfusion is based on the amount of blood delivered to the tissue per unit time per unit tissue mass ([ml/ml/s] or [ml/100 g/min]). The methods described in this section characterize perfusion mostly indirectly through its proxies, i.e., blood oxygen saturation or skin temperature. For example, blood oxygen saturation is the balance between blood supply (perfusion) and demand (consumption). Thus, one can indirectly assess the blood perfusion by determining blood oxygen saturation in the microvasculature.

5.1.1 Laser Doppler

Photons change frequency if mobile scatterers are illuminated with coherent light (the Doppler effect). Red Blood Cells (RBCs) are primary mobile scatterers in cutaneous circulation; thus, the Doppler effect can be used to evaluate their velocities.

Laser Doppler Flowmetry (LDF) is a class of noninvasive techniques that utilize the optical Doppler effect to assess microcirculatory blood perfusion. In practice, LDF often refers to a single-point measurement with a fiberoptic (contact) probe (experimental modality), while Laser Doppler (Perfusion) Imaging or LDPI refers to a clinical imaging modality.

Laser Doppler Imaging is a relatively mature diagnostic modality used in many clinical situations. It is primarily used to assess burn depth but is also applied in surgery, wound healing, and general vascular diagnostics. However, it has very limited applications in diabetic wound care.

In a comparison study,[6] Laser Doppler Flowmetry and Transcutaneous Pressure of Oxygen (TcPO$_2$) were performed in 25 patients with chronic lower limb ulcers with various etiology (experimental arm) and 25 healthy individuals (control arm). A statistically significant difference ($p<0.05$) was found between the LDF values of the two groups. However, no statistically significant differences were found between the two groups via the TcPO$_2$ measurements.

One study[7] used LDPI to measure the circulation in ischemic ulcers. Ten out of the 25 included patients were diabetic foot patients. They found that changes in the circulation measured by LDPI may coincide with changes in the number of visible capillaries within an ischemic ulcer.

Some studies used LDPI to measure the effect of different healing techniques for diabetic foot ulcers and showed that LDPI could measure the circulation in Diabetic Foot Ulcers and that this technique could be used to assess microcirculation.[8]

5.1.1.1 Laser Doppler: Practical Considerations

It is recommended[9] that measurements are taken under the same conditions where appropriate (e.g., same site studied, same ambient temperature, complete acclimatization, caffeine-, nicotine- and vasodilator-free, etc.), with the same LDPI parameters (e.g., wavelength, scanning speed, scanning distance, DC values, and image normalization) to ensure valid comparisons.

In addition, temporal variations in blood flow over short periods (days to weeks)[10] and more extended periods (due to seasonal changes)[11] may need to be taken into account, as does the effect of temperature variation, physical and mental activity and consuming certain vasoactive substances (for example, caffeine), which have been shown to have a significant impact on the laser Doppler technique.[11]

5.1.2 Laser Speckle Imaging

A speckle pattern is formed by reflecting coherent light from a rough surface or by reflecting or transmitting the light through a medium with refractive index distribution. This phenomenon results from the interference of different reflected portions of the incident beam with random relative optical phases. Determinations of RBC velocity may be obtained by assessing the temporal statistical behavior of speckles.

Laser Speckle (Perfusion or Contrast) Imaging, LSPI, or LSCI, is a relatively mature diagnostic modality used in many clinical situations, including the noninvasive assessment of blood flow of cutaneous wounds.[12] However, it has very limited applications in diabetic wound care.

In,[13] Laser Speckle Imaging was used to study whether iontophoresis increases skin microcirculation in diabetic patients (iontophoresis is a process of transdermal drug delivery). Typically, a vasodilator such as acetylcholine was used to measure a tissue's 'microvascular capacity.' In particular, the authors found that LSCI could be used to diagnose small changes in microcirculation in specific areas, such as the skin on the foot or ankle.

5.1.3 Spectroscopic Methods

Light in the visible and Near-Infrared (NIR) ranges of the spectrum delivered to biological tissue undergoes multiple scattering and absorption events. Hemoglobin, water, fat, and melanin are the primary absorbers in the tissue. It is well established that this reflected light carries quantitative information about tissue pathology.

Spectroscopic techniques refer to a broad range of technologies that use spectroscopic principles to extract information about underlying tissue.

Near-Infrared Spectroscopy (NIRS) typically refers to a single-point contact measurement, e.g., using a fiber probe.

In recent years the progress in cameras, image analysis techniques, and computational power have made it possible to extend the advances in biospectroscopy into clinical imaging modalities.

Biomedical Hyperspectral Imaging (HSI) aims to record the spectrum for each image pixel and extract the concentration of tissue chromophores. Thus, hyperspectral imaging is the natural extension of color (RGB) imaging and biospectroscopy (a single-point measurement). The spectrum at each pixel as a function of a wavelength (λ, nm) can be considered a spectroscopic input, which can be decomposed, and spectral signatures can be found.

5.1.3.1 Near-Infrared Spectroscopy (NIRS)

Near-Infrared Spectroscopy (NIRS) typically refers to contact modalities that provide single-point measurements. NIRS modalities are usually based on the principle of spatially resolved spectroscopy.

Based on spectral parameter measurement in animal experiments with rats, it was found that the diabetic group has a significantly higher average value of absorption coefficient (~100%) at wavelength 780 nm compared to a healthy group during the wound healing process. Compared to healthy rats, diabetic rats have a higher (by ~ 30%) value of reduced scattering coefficient.[14]

A clinical study deals with the application of diffuse reflectance intensity ratios based on oxyhemoglobin (HbO_2) bands (R542/R580), ratios of oxy- and deoxyhemoglobin (RHb) bands (R580/R555), total Hemoglobin (tHb) concentration, and hemoglobin Oxygen Saturation (SO_2) between normal and Diabetic Foot Ulcer sites. Considerable differences in RHb and HbO_2 were observed for ulcer regions compared to control sites in clinical studies. In addition, the total Hemoglobin (tHb) concentration was higher in foot ulcers.[15] The authors also observed higher values of SO_2 during the healing process, which follows from an increase in blood supply and, in due course, SO_2 reduces as the wound heals.[16]

Biospectroscopy can be used for other clinical indications. Poosapadari et al.[17] used spectroscopy to discriminate between the two most prevalent pathogens in Diabetic Foot Ulcer (DFU) patients (*S. aureus* and *E.*

coli), with 100% sensitivity and 75% specificity in detecting the presence of these infections and a 100% negative predicted value in excluding the infection in such wounds.

5.1.3.2 Hyperspectral and Multispectral Imaging

Hyperspectral and multispectral imaging refers to a broad group of non-contact spectroscopy-based technologies that extract information about underlying tissues from relatively large areas.

Often, the terms 'hyperspectral' (HSI) and 'multispectral' (MSI) imaging are used interchangeably. The subtle distinction between them is based on an arbitrary 'number of bands.' Multispectral imaging deals with several images at discrete and somewhat narrow bands placed at particular spectral points (e.g., isosbestic or absorption maxima). Thus, multispectral images do not produce the 'spectrum' of an object but rather sample the spectrum at several points. On the other hand, hyperspectral imaging implies narrow, equally spaced spectral bands over a continuous spectral range, which can be considered a spectrum.

In biomedical applications, hyperspectral or multispectral imaging is used primarily to extract data about components of the blood, which are chromophores in the visible and NIR spectrum.[18] The primary outcome of HSI/MSI methods is tissue Oxygen Saturation (SO_2) maps, which indicate abnormalities in blood circulation. Occasionally, oxyhemoglobin (HbO_2), deoxyhemoglobin (RHb), and total Hemoglobin (tHb) or their proxies are also reported.

HSI/MSI is increasingly being used within different clinical diagnostic areas. In line with many reported clinical applications [64], hyperspectral imaging demonstrated its utility in wound care in general and DFU in particular. For example, hyperspectral imaging has been used to assess tissue viability and health in diabetes patients at risk of foot ulceration.[19,20] The applications of HSI/MSI in wound care were reviewed in.[21] Here, we briefly present significant findings for DFU.

Greenman et al.[22] used HSI to measure blood oxygen saturation in healthy, diabetic, and diabetic neuropathic individuals. They found that in the foot at resting, SO_2 was lower in the neuropathic group (30±12; p=0.027) than in the control (38±22) and non-neuropathic groups (37±12). In Figure 5.2, one can see an example of oxygen content visualization using HSI/MSI.

Serena et al.[23] compared SO_2 assessed by multispectral imaging with $TcPO_2$ measurements in 12 locations in 10 patients. They found that $TcPO_2$

54 ■ Optical Methods for Managing the Diabetic Foot

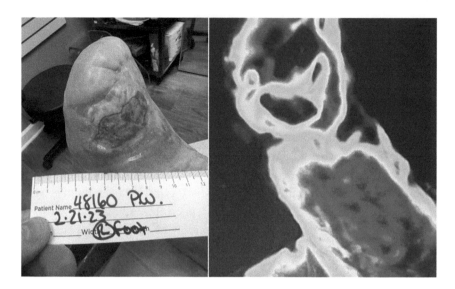

FIGURE 5.2   HSI oxygen content imaging. A Diabetic Foot Ulcer was assessed using HSI imaging (right panel). A false-coloring of the image shows low oxygen content (blue and green hues) in the wound bed compared to the surrounding tissue (red). Interestingly, the peri-wound area exhibits areas of increased oxygen content, probably due to local hyperemia to the region (asterisk), and areas of lower oxygen content where skin breakdown is visible (arrows). (Image courtesy of Dr. Matthew Regulsky.)

provides higher readings in general. However, SO_2 extracted by the multispectral imaging system and $TcPO_2$ are strongly correlated ($r=0.74$).

5.1.3.2.1   Blood Oxygen Saturation and Wound Healing

Khaodhiar et al.[70] measured RHb and HbO_2 in nondiabetic controls ($n=14$) as well as diabetic patients with ($n=10$) and without DFU ($n=13$) four times over 6 months using HSI. HbO_2 and RHb measurements in the peri-wound of non-healing ulcers were lower than in healing ulcers ($p<0.01$) and contralateral foot ($p<0.001$).

Nouvong et al.[71] measured HbO_2 and RHb in diabetic patients ($n=44$) over 24 weeks using HSI. They found that in the peri-wound area, HbO_2 and SO_2 were higher around DFUs that healed (85 ± 21 and $66\pm9\%$, respectively) vs. non-healing (64 ± 22 and $60\pm10\%$, respectively).

Jeffcoate et al.[24] compared SO_2 measurements using HSI against blood gas analysis in blood samples (in vitro) in patients with DFUs ($n=43$). They found a negative association between SO_2 and healing by 12 weeks ($p = 0.009$) and a significant positive correlation between oxygenation assessed by HSI and time to healing ($p = 0.03$). In addition, a strong correlation between SO_2 and blood gas measurements ($r = 0.994$) was found.

A retrospective clinical study was conducted to determine if MSI/HSI technology can be used to evaluate wounds and adjacent soft tissues, identify patterns involved in tissue oxygenation and wound healing, and predict which wounds may or may not heal. Most wounds progressing toward healing showed a wispy and gradual decrease in SO_2 as the distance from the wound increased. Landsman[25] described this as having an appearance resembling a ray of sunlight, with the central red region moving to a combination of yellow and red as the distance away from the wound increased. This was probably the most predictive sign that the wound was starting to heal. An example of a positive response to hyperbaric oxygen therapy in a patient with arterial insufficiency is depicted in Figure 5.3.

5.1.3.2.2 Prediction of DFU Development

Yudovsky et al.[26] using dataset,[66] assessed the utility of HSI in predicting DFU development. Diabetic patients ($n=54$) at risk of DFU were monitored and evaluated retrospectively for ulceration. They found that the difference in HbO_2 (and RHb at a minor degree) value between wound and peri-wound was a predictor of ulceration. The authors constructed an index that could predict tissue at risk of ulceration with a sensitivity and specificity of 95% and 80%, respectively, for images taken, on average, 58 days before skin breakdown.

5.1.3.2.3 Other Clinical Applications

Establishing optical biomarkers that predict healing has value in clinical practice as a guide to planning therapy. There are numerous therapies for the treatment of Diabetic Foot Ulcers. The associated costs range from hundreds of dollars to tens of thousands of dollars. Clinically, using expensive therapies for patients with a high likelihood of healing with basic care would be inappropriate.

Yudovsky et al.[27] demonstrated the utility of biospectroscopy and HSI/MSI to determine the thickness of the epidermis layer.

Several studies demonstrated the utility of HSI/MSI to detect tissue water content, which can be helpful for several clinical indications, including deep tissue injury and preclinical edema identification. It also can have relevance to the diabetic population. For example, in,[28] the authors found differences in skin water content between diabetic and nondiabetic populations. In,[29] the authors used the 1400–1500nm band to extract the

56 ■ Optical Methods for Managing the Diabetic Foot

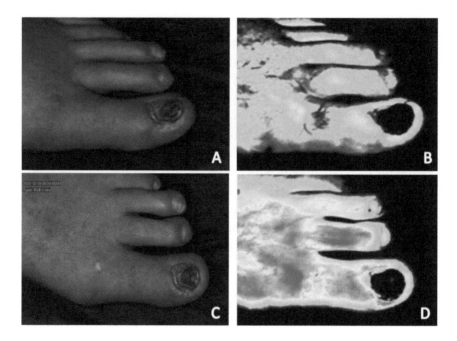

FIGURE 5.3 A positive response to hyperbaric oxygen therapy in a patient with arterial insufficiency. (A) Baseline clinical photograph; and (B) Near-Infrared Image assessment of the left dorsal foot in a patient with arterial insufficiency referred by hyperbaric oxygen therapy. Note reduced signal to the dorsal foot (C) immediately prior to a single session of hyperbaric oxygen therapy and immediately following (D) this treatment. (Reproduced from under CC BY 4.0 license.) J Arnold,VL Marmolejo,"Interpretation of near-infrared imaging in acute and chronic wound care", *Diagnostics*,11:778 (2021).

skin's hydration. In,[30] this approach was extended to the 970nm range, which allows deeper light penetration into the tissue.

5.1.3.2.4 HSI/MSI: Practical Considerations

There is growing evidence that the epidermal layer can impact the extraction of physiological parameters in a number of ways.

It is known that skin tone plays an important role. In particular, it is expected that HSI/MSI can extract physiological parameters only from medium and lightly-pigmented skin (skin tone I–IV).

Epidermal thickness also plays an important role. In particular, it has been shown[31] that the dermis almost does not contribute to reflected signal for stratum corneum thickness of 1.5mm and more. Thus, interpreting physiological parameters extracted from areas with thick epidermal layers, e.g., calluses, should be taken cautiously.

5.1.3.2.5 HSI/MSI: Methodological Aspects

Because various methods, protocols, and study populations are being used, the literature often reports controversial measurements and conclusions.

Several critical factors may impact study results. One can be attributed to the HSI wavelength range. Currently, there are two different HSI/MSI approaches to measuring oxyhemoglobin, which are based on the incident light range. One method uses light in the visible spectrum. The other is to use light in the Near-Infrared (NIR) range. The longer NIR wavelength provides deeper penetration, which may be seen as a benefit until one considers the vascular anatomy (see Figure 5.4).The visible light method only inspects the most superficial papillary blood supply, which supplies the wound margins. In other words, it is a very discrete measurement of a single anatomic layer.

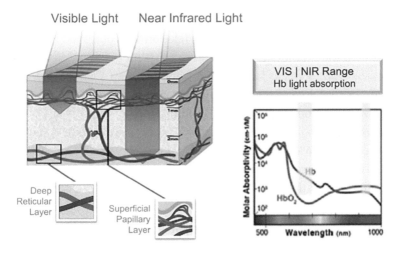

FIGURE 5.4 Oxyhemoglobin and deoxyhemoglobin wavelength absorption curves can be distinguished in visible and Near-Infrared Ranges, as highlighted in grey. Visible light spectroscopy measures only the superficial papillary layer associated with healing granulation tissue. NIR spectroscopy penetrates deeper and is a combined measurement of the reticular and papillary systems. (Modified from with permissions.) C Weinkauf, A Mazhar, K Vaishnav, et al. "Near-instant noninvasive optical imaging of tissue perfusion for vascular assessment", *J Vasc Surg*, 69(2):555–562 (2019).

In contrast, the NIR approach measures two layers: the superficial papillary layer and the deep reticular layer. These two layers have different functions, and the combined measurement of oxyhemoglobin in two layers is more difficult to interpret. This may partially explain the variability of NIR-based measurements and the interpretation of these measurements as healing predictors (see Figure 5.4).

Other differences can be attributed to the different clinical methodologies. For example, a visible light-based spectral study of patients with Wagner Grade I–II DFUs not complicated by severe Peripheral Artery Disease (PAD) and/or infections found that oxyhemoglobin (oxyHb) measurements were a reliable predictor of healing.[32] Post-debridement hyperspectral images were taken to evaluate the ulcer size, peri-wound oxyHb level, deoxyhemoglobin (DeoxyHb) level, and Oxygen Saturation (O_2Sat) for four consecutive visits. Twenty-seven patients were followed, out of whom seven healed their DFU while the remaining 20 failed to heal their DFU. The average time between each visit was 3 weeks.

Patients with lower oxyHb measurements had a higher probability of healing their DFU, which reached statistical significance during the second visit. A significant inverse correlation was observed between the oxyHb and the percentage of DFU size reduction between Visit 1 and the end of the study 9 weeks later at Visit 4 and between Visit 2 and Visit 4.

A significant inverse relationship was observed for the percentage of ulcer size reduction between Visit 2 and the end of the study at Visit 4 and Visit 2 oxyHb ($r = -0.65$, $p = 0.001$). For Visit 2, oxyHb had 85% sensitivity, 85% specificity, 66% positive predictive value, and 94% negative predictive value. The negative predictive value is notable because it identifies patients who will not heal their DFU and need more expensive/intensive therapies such as bioengineered skin products.

However, previous studies using visible light-based HSI imaging found the opposite oxyHb relationship. Namely, high oxyHb was associated with an increased probability of DFU healing, opposite to the findings of the Kounas study.[33,34] Nevertheless, a similar result of an inverse association between oxyHb and DFU healing over 12 weeks was observed in another HSI study by Jeffcoat.[35]

The most probable explanation for these diverging results lies in different clinical methodologies. In previous studies, imaging was performed before ulcer debridement. In contrast, the Kounas study performed imaging after debridement. Post-debridement imaging provides more discrete measurements of healthy peri-wound tissue close to the granulating tissue and free of any callus that interferes with the spectral signature of oxyHb. The lower oxyHb measurements indicate that oxygen consumption is increased in this tissue. The high oxyHb values observed in previous studies are likely related to the fact that the tissue was not very close to the granulating base of the ulcer, and the tissue is less metabolically active. Consideration should also be given to the arteriovenous shunting associated with neuropathy. In particular, arteriovenous shunts lead to reduced oxygen consumption and higher oxyHb measurements.

Another important shortcoming of various HSI/MSI studies is that they often report HbO_2 and RHb in some arbitrary units, which don't allow direct interpretation or comparison. Potentially, it can be improved using a so-called 'physiological' calibration. In this case, a calibration image is taken on the patient's unaffected skin or mucosa area (e.g., eyelid).

5.1.4 Spatial Frequency Domain Imaging (SFDI)

Tromberg et al. in[36] developed rapid, noncontact imaging for quantitative, wide-field characterization of turbid media's optical absorption and scattering properties based on Spatial Frequency Domain Imaging (SFDI). This group experimentally demonstrated that by projecting sinusoidal light patterns onto tissue, one could determine the tissue's optical properties by measuring the relative decay of spatial patterns of differing frequencies. An algorithm was proposed for reconstructing 3-D images directly from measurements made by illuminating tissue with sinusoidal patterns[37] and experimentally proven as a quantitative optical tomography of subsurface heterogeneities (e.g., blood vessels) using spatially modulated structured light[38] and collecting images separated in terms of absorption and reduced scattering coefficients.[39] SFDI has been demonstrated preclinically to track wound healing in a diabetes model.[40] A clinical study has applied SFDI to measure lower extremity perfusion to predict both the healing and formation of Diabetic Foot Ulcers.[41,42]

5.1.5 Thermography

Thermography has been tried in wound care since a long time. In recent years, it has been adopted into clinical practice. There are two primary temperature monitoring modalities in diabetic wound care: noncontact (thermal camera) and contact (temperature monitoring mat and temperature monitoring wearables).[43]

This book considers thermal camera applications as a noncontact imaging modality. Thermal imaging sensors operate in the middle (MWIR, 3–5 μm) or long (LWIR, 8–14 μm) wavelength Infrared (IR) spectral ranges. As such, noncontact thermography is often referred to as Infrared thermography or IRT.

The most common materials for thermosensors are amorphous Si (a-Si), InSb, InGaAs, HgCdTe, and Quantum Well-Integrated Photodetectors (QWIP) arranged into focal plane arrays or FPA. While room temperature operations are sufficient for many applications, the cryogenic cooling of IR detectors is required for specific applications to achieve high performance. Uncooled detectors are mainly based on pyroelectric and ferroelectric materials or microbolometer technology. Thermal imaging sensors for biomedical research and clinical applications use low-cost, uncooled FPA microbolometers in the LWIR range. The resolution of thermographic sensors is much lower than regular RGB cameras. Currently, it is in the range of 60x80 or 120x160 for standard applications and 640x480 for high-end applications.

The utility of thermography in wound care has been known since the early 1960s when Lawson et al.[44] used Infrared scanning to predict burn depth with an accuracy of 90%, as confirmed by histology.

Thermography has significant potential as an adjuvant technique in diabetic foot assessment. For example, elevated temperature is a reliable marker of inflammation and can thus predict the risk of ulceration, infection, and amputation.[45] Similarly, the decreased temperature may be a sign of insufficient blood supply and indicate ischemia (see Figure 5.5). The applications of thermography in wound care were reviewed in.[46]

Various studies have investigated IRT's role as a diagnostic tool for detecting Peripheral Arterial Disease (PAD) and as a tool for post-procedural assessment/surveillance of angioplasties. Under controlled conditions, skin temperature directly reflects arterial blood flow, which transports oxygen, nutrients, and heat to the extremities.[47]

Although speed, convenience, and reproducibility hold promise, there is no consistency in the measurement protocols, measurement locations, or measurement analysis. Locations ranged from the anterior tibia and ankles, the toes, the dorsum of the foot, and the plantar side– the most common site across all study methods. Table 5.1 provides a summary of study types and locations.

5.1.5.1 Diagnosis of PAD

The extremities' high surface-to-volume ratio means the extremities will cool if there is a decrease in blood flow. Arteries serve as heat conduits, and

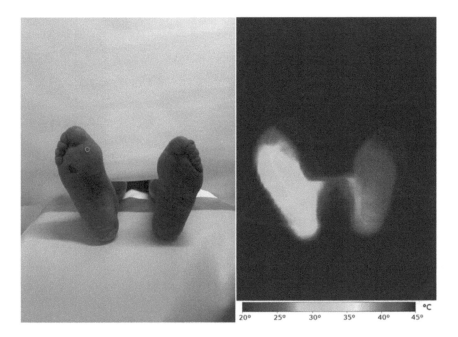

FIGURE 5.5 Thermographic assessment of Peripheral Arterial Disease (PAD). Limbs with PAD show significantly lower temperatures than the contralateral extremity. In this case, a diabetic individual with PAD on the left foot is presented. A screen is used to thermally isolate the rest of the body, showing a stark contrast between the feet temperature.

a reduction in blood flow from PAD typically leads to a drop in skin temperature. However, using absolute temperatures is problematic as there is enormous variability in skin temperatures from hour to hour, gender differences, hormonal differences, and autonomic tone.

A healthy person has a nearly constant human core temperature (within +0.6°C), which can be used for routine monitoring. Because the skin plays an essential role in thermoregulation, it serves as an effective heat radiator or insulator based on cutaneous blood flow. In a vasoconstricted state, it approaches the insulation value of cork (see Figure 5.6). In contrast to core temperature, skin temperatures can change substantially over a large range.

5.1.5.2 Thermal Asymmetry

Because skin temperatures are highly variable between individuals, absolute temperatures have little value. What has proved to be useful is the use of *asymmetry values* where a Region of Interest (ROI)

TABLE 5.1 Summary of Clinical Thermographic Studies

Author (Year)	Design	Study Objective	Location	Ref
Bagavathiappan (2008)	Case report	PAD Detection	Shins, Feet	48
Philip (2009)	Case report	PAD Detection	Shins, Feet	49
Carabott (2021)	Observational	PAD Detection Limb Elevation	Feet	50
Chang (2020)	Observational	Revascularization	Feet	51
de Carvalho Abreu (2022)	Nonrandomized	ABI Correlation	Feet	52
Gatt (2018)	Nonrandomized	PAD Detection	Feet	53
Gatt (2018)	Nonrandomized	DFU complications	Feet	54
Hosaki (2002)	Case report	PAD Detection	Feet	55
Huang (2011)	Observational	PAD Detection	Shins, Feet	56
Ilo (2020)	Observational	PAD Detection	Feet	57
Ilo (2021)	Nonrandomized	Revascularization	Feet	58
Renero-Carrillo (2021)	Nonrandomized	Revascularization	Feet	59
Staffa (2016)	Nonrandomized	Revascularization	Feet	60
Wallace (2018)	Observational	PAD Detection	Feet	61
Wang (2004)	Observational	PAD Detection	Ankle, Feet	62
Zenunaj (2021)	Observational	Revascularization	Ankle, Feet	63

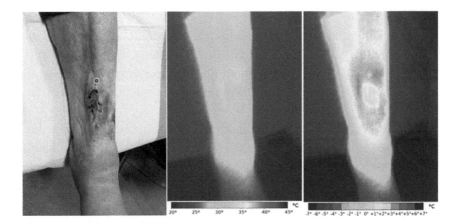

FIGURE 5.6 Use of absolute vs. relative temperature gradients for infrared thermal imaging. These panels show a traumatic ulcer in a diabetic individual. The middle panel reports the absolute temperature distribution in °C, while the right one is the temperature relative to the circle on the middle panel (reference point). By using a temperature gradient, differences in the wound bed and peri-wound area are maximized; while confounding factors are adjusted.

is compared to another region on the patient, or perhaps 'pre-intervention' values are compared to 'post-intervention' values on the same patient. Caution should be exercised when using IRT as a perfusion screen when there is local hyperemia from infection or reperfusion. The presence of inflammatory changes will lead to an overestimation of perfusion.

An illustration of the thermal asymmetry concept is the *'side-to-side'* temperature differences *between* feet are greater in patients with diabetes than among healthy controls, where there is a little asymmetry between feet. Skin temperature usually shows symmetrical bilateral distribution; thus, *strong asymmetry* indicates an abnormality that can serve as a screening marker for more advanced tests. Skin temperature variations over 2.0°C may be useful for identifying pathological situations in the diabetic foot.[59]

For patients with diabetes, mean temperatures significantly differed between the subgroups with angiopathy, neuropathy, and neuroischemia. Notably, neuropathic feet were warmest, followed by neuroischemia and then angiopathy without neuropathy. Patients with DM with neuropathy (27%) had a higher mean temperature compared to the healthy controls (plantar side: 29.4°C vs. 27.7°C ($p = 0.001$).[59]

5.1.5.3 Thermal ABI

Using the thermal asymmetry concept, Wallace et al. developed the Thermal ABI (tABI).[63] The tABI is calculated using the highest plantar temperature and dividing it by the highest temperature obtained in the bilateral hands. This method is very similar to the calculation of the traditional ABI using the highest systolic pressure.

Validation of a new measurement method for application to medical practice requires comparison with gold standard techniques. The Bland-Altman analysis is frequently applied in studies investigating the agreement between two different measurement methods. Thus, ABI and tABI were compared using the Pearson correlation and Bland-Altman analysis. The Bland Altman plot was used to determine the agreement between ABI and tABI.

A positive correlation was found between ABI and plantar temperature ($r = 0.66$, $p = 0.001$). A positive correlation was found between ABI and tABI ($r = 0.83$, $p < 0.0001$). The Bland-Altman analysis also showed agreement between ABI and tABI (bias −0.01, 95% limit of agreement −0.13 to −0.12).

The tABI may be useful in patients with calcified tibial vessels, such as diabetics and renal failure patients, although this has not been evaluated. The tABI may be more accurate given the limitations of traditional ABI with calcified vessels, particularly given the promising findings correlating Doppler sensors with digital thermal studies comparing toe–finger thermography with ABI.[64]

These findings are consistent with Abreu et al.[54] who correlated IRT findings to ABI measurements in patients with PAD. IRT and ABI in patients with noncalcified arteries showed a strong direct relationship. The Spearman correlation between ABI and the mean plantar temperature had a significant correlation ($r=0.7$) in patients without arterial calcification.

5.1.5.4 Predictors for Healing and Major Amputation

Critical Limb Ischemia (CLI) is the most advanced form of Peripheral Artery Disease (PAD). Current guidelines consider Endovascular Therapy (EVT) an acceptable treatment for CLI attributable to infra-popliteal lesions, the most common distribution in diabetes. However, in this high-risk population, EVT does not have predictable success. Post-EVT for infra-inguinal lesions, 43% of patients will require a reintervention, 37% will require amputations, and 43% will have wound recurrence. The need for a better EVT outcome predictor is needed.[65]

Currently, available tools for assessing EVT treatment outcomes include the Ankle-Brachial Index (ABI), toe pressure, Duplex ultrasound, skin perfusion pressure, and Transcutaneous Pressure of Oxygen (TcPO$_2$) measurements have various limitations of times and were discussed earlier in this book.

Limb thermography with thermal measurements reflects blood flow secondary to vessel patency and is a reasonable surrogate for EVT outcome. A decrease in temperature may indicate the presence of arterial occlusive disease or early re-occlusion.

Chang et al. hypothesized that a change in foot temperature after EVT may be a predictor for wound healing and freedom from major amputation.[53] Thermal images of each foot were divided into five zones; one on the dorsal and four on the plantar aspect, corresponding to the angiosomes in the foot. A pre-EVT Thermal Asymmetry Index (TAI) was defined as the lowest foot temperature minus the average foot temperature. The average temperature was calculated by adding the average temperature for each of the five zones and dividing it by five to produce the average temperature for the entire foot. A post-EVT TAI was calculated using the same method.

The average temperature gain post-EVT was not significantly different between those with good or poor outcomes, which reinforces the concept that absolute skin temperatures are rarely useful because of the high variability between patients. However, creating an asymmetry index was useful. The post-EVT TAI was significantly higher for patients with a better outcome. The best cutoff value of post-EVT TAI for predicting freedom from major amputation was determined by ROC curve analysis. A post-EVT TAI $\geq 1.3°C$ predicted freedom from major amputation during a 6-month follow-up, with a sensitivity of 71%, specificity of 54%, positive predicting value of 60%, and a negative predicting value of 65%, ($p= 0.004$).

5.1.5.5 Dynamic Thermography

Dynamic Thermography (DT),which uses measurements before and after a physiologic challenge, is a useful clinical tool. In particular, the 6-minute walking test (6MWT) is an established physiologic challenge to evaluate PAD. Huang et al. investigated the IRT measurements of the anterior shin and plantar foot before and after the 6MWT and the correlation with ABI measurements.[58]

The exercise-induced Temperature Change (eTC) was defined as the post-exercise temperature minus the pre-exercise temperature. The resting temperatures were similar in PAD and non-PAD patients. However, the post-exercise temperature dropped in the lower extremities with arterial stenosis, but was maintained or elevated slightly in the extremities with patent arteries. The temperature changes in PAD vs. non-PAD patients were 1.25°C vs. 0.15°C ($p<0.001$). The eTC changes at the sole positively correlated with the 6MWT (Spearman correlation coefficient $r= 0.31$ ($p<0.03$), and it correlated with ABI measurements (Spearman correlation coefficient $r= 0.48$ ($p<0.001$).

5.1.5.5.1 Cooling Challenge for PAD
Pakarinen et al. explored the value of Dynamic Thermography using thermal and hydrostatic modulation in combination with the angiosome concept, where the lower limb is divided into regions supplied by a single artery.[66] The hydrostatic modulation test was a derivation from the Ratschow test, in which the lower limbs are passively elevated and lowered to induce active ischemia and congestion.[67,68] Thermal modulation was performed by simultaneously cooling both feet using moldable cooling pads. Alternative cooling methods, such as cold-water submerging[69] or cooled air,[70] were found unsuitable for CLI patients. The thermal cameras followed the cooling process until a 15% temperature drop from the initial reading for both limbs was achieved, corresponding to approximately 2.5–5°C.

The hydrostatic modulation test was divided into three subphases: stabilization (8 min), limb elevation (3 min), and recovery (3 min). There was no detectable relation between hydrostatic test phases and mean angiosome temperatures.

In contrast, the thermal modulation test augmented the thermoregulation deficiencies, and high correlations was found with all reference metrics. The thermal recovery time was higher for the PAD (88%) and CLI (83%) groups compared to healthy subjects. The contralateral symmetry was high for the healthy group and low for the CLI group. The thermal recovery time negatively correlated to TBI ($r = 0.73$) and ABI ($r = 0.60$).

5.1.5.6 Flap Perfusion
Other investigators have demonstrated the value of Dynamic Thermal Analysis (DTA) using a cooling challenge to accentuate or magnify small temperature differences associated with flap perfusion, which are not apparent under normal conditions. An example of DTA is used in plastic surgery to identify and monitor arterial perfusion of free flaps.[71]

For selected chronic wounds, a perforator-free flap for coverage of an ulcer is useful. A prerequisite to flap survival is the identification of the dominant perforator artery which supplies the flap. A 'freestyle' approach based on general anatomy is often used to design these flaps, as conventional imaging techniques for specific perforator identification may be too expensive or unavailable. However, unexpected variations in anatomy can lead to flap failure. The use of a thermal imaging camera is a cheaper and more universal means to identify the requisite perforators upon which a free flap can be designed and monitored.

In particular, thermography (including smartphone thermography) can be used preoperatively to identify arterial perforators or vascular 'hot spots' within the desired donor site territory. Intraoperative selection of the arterial perforators and flap dissection can be facilitated using IRT as a real-time guide. In addition, intermittent postoperative monitoring using thermal images provides a comparison method to determine if the anastomotic perfusion is compromised.

Smartphones provide a lower resolution image than the more expensive professional cameras.[72] The smartphone IRT camera is less sensitive, so initial thermal stress or 'cooling challenge' is required. This is why Dynamic Thermography (DT) using a physiologic challenge is a preferred method.[73,74,75] DT is most simply performed using Muntean's method of spraying the flap donor site with isopropyl alcohol followed by thermography to identify the precise location of flap perforators.[76]

A concordance study compared smartphone perforator detection with CT angiography.[77] Smartphone IRT was highly accurate, with a sensitivity of 100% and a specificity of 98%. Other recognized techniques for perforator identification include magnetic resonance angiography or color Duplex ultrasound, which are reliable alternatives but are expensive, time-consuming, and not universally available.[78,79]

Thermography offers immediate insights to provide adequate intraoperative confirmation of perfusion. The thermogram also provides an additional means for postoperative monitoring. Surveillance is important in the immediate postoperative period as there can be complications such as arterial kinking, compression from edema or minor hematomas, etc. Early identification leads to early intervention and successful outcomes. A thermogram is a near-perfect surveillance tool, simple to obtain, noninvasive, and accurate. However, it is not continuous, and some interpretation of the flap temperature is required. Where there is bedside uncertainty with nursing personnel, the same smartphone used to make the thermogram can be used to send these pictures wherever needed for a second opinion.

5.1.5.7 Angioplasty Assessment

A prospective study compared skin temperature changes in the feet before and after revascularization. The majority of patients had intermittent claudication. IRT was performed on five plantar sites and five dorsal sites of each foot for 10 sites per foot.[60]

Revascularization methods included Percutaneous Transluminal Angioplasty (PTA), intra-arterial stent implantation, peripheral bypass surgery, and femoral thromboendarterectomy. After the operation, ABI, TP, and TBI increased significantly. Patients with an apparent decrease in these measurement values lost their revascularization result. In the revascularized legs, the mean change before and after the procedure was statistically significant in the ABI (+0.28 $p<0.0001$), TP (+21.6 mmHg $p<0.0001$), and TBI (+0.14 $p<0.0001$). The mean temperature increased post-procedure +0.5°C on the plantar side ($p = 0.02$) and +0.6°C on the dorsal side ($p = 0.03$).

A similar study compared IRT changes in the foot following infrafemoral endovascular procedures.[65] Using four plantar measurements per foot, a statistically significant asymmetry relationship was found with the IRT measurements at each point of the ischemic and contralateral feet. In this study, the majority of patients had Critical Limb Ischemia. The mean difference between the two limbs at the foot was 1.7°C which is sixfold higher than those reported in the literature for patients with intermittent claudication (0.3°C).[59] This finding suggests the larger the asymmetry, the more severe the disease.

Skin temperature normally shows symmetrical distribution with the contralateral limb; thus, strong asymmetry indicates an abnormality.[80,81,82] Skin temperature variations over 2.0°C are useful for identifying pathological situations in the diabetic foot.[83,84,85] In revascularization studies, the mean side-to-side difference between feet was higher before revascularization and diminished after the procedure (see Figure 5.7).

The temperature of the feet in patients with diabetes was higher than in patients with PAD or healthy controls. Neuropathy and neuroischemia raise the skin temperature of the foot in contrast to atherosclerotic ischemia without diabetic neuropathy. One should also consider that the presence of any wound, even without infection, also increases skin temperature.[60]

Optical Diagnostic Techniques ▪ 69

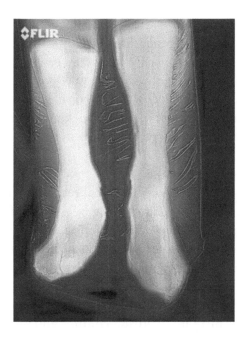

FIGURE 5.7 Post-angioplasty thermal changes. The right leg of this patient underwent balloon angioplasty. Seventy-two hours after the procedure, a significant increase in the temperature of the leg could be observed compared to the contralateral one. This indicates reactive hyperemia to the newly patent vessel and has been demonstrated to be predictive of the long-term success of the procedure. (Photograph courtesy of Dr. Jesus E. Arriaga-Caballero.)

5.1.5.8 Diabetic Neuropathy and IRT Measurements

Elevated skin temperatures are due to microvascular dysfunction. The dermal microcirculation blood flow comprises the nutritional capillaries and thermoregulatory Arteriovenous (AV) shunts. Glabrous skin, skin without hair, is found on the soles of the feet and the palms of the hands. This skin gets 25% of its blood flow from nutritional capillaries and 75% from AV shunts. Non-glabrous skin does not possess AV shunts, and blood flow is largely nutritional capillaries. Endothelial dysfunction in small vessels is an early marker of vascular disease. Microvascular dysfunction is a systemic process where symptoms of PAD are caused by a lack of blood flow in the nutritional capillaries that does not meet the metabolic demand of the tissues. When blood flow improves after revascularization, the need for AV shunting decreases.[86,87,88]

This thermal pivot can be utilized in evaluating wound healing, as seen in the following example. There were two patients with diabetic neuropathy but no wounds and two patients without diabetes and also without wounds.[60] Both groups had PAD and received transluminal angioplasties. In both groups, the temperature asymmetry decreased. This change occurs with improved blood flow post-transluminal angioplasty. Deep nutritive blood flow turns normal after revascularization and decreases shunting (heat transfer) through microvascular AV fistulas. Correspondingly, the skin temperature decreases.

5.1.5.9 Inflammation Detection

Inflammation is a complex physiological response triggered by tissue damage. Inflammation is characterized by a series of events aimed at removing the injurious agent and initiating tissue repair. One of the earliest responses includes the induction of hyperemia, which refers to a localized increase in the blood flow to the affected area, leading to the cardinal signs of erythema and warmth.[89] The process by which acute inflammation induces hyperemia and an increased thermal load includes histamine, prostaglandins, and bradykinin-induced meta-arteriole vasodilation; increased vascular permeability, which also leads to vascular fluid leakage and edema; and decreased vascular tone in the capillary bed, also leading to increased flow and plasma leakage.[90]

Infrared thermography is an excellent tool for visualizing acute and chronic inflammatory changes.[48] For example, localized temperature increases after tissular injury were documented.[91] The authors followed a series of patients experiencing traumatic injuries to the extremities. They obtained time-lapsed thermal images showing dramatic increases in the affected areas' heat load and dissipation of the heat as time passed. Interestingly, in one case, the heat load followed the vascular distribution of the hand, demonstrating how flow and heat are intimately related during inflammation. Similarly, in chronic inflammatory conditions, localized temperature increases have been demonstrated to correlate with areas of inflammation diagnosed by other imaging modalities, such as ultrasound,[92] or to clinical severity scoring systems.[93]

In the case of the assessment of diabetes ulceration risk, several approaches based on plantar temperature distributions have been proposed. Benbow et al.[94] assessed the risk of ulceration and ischemic foot disease based on the mean foot temperature determined from eight standard sites on the plantar surface. They found that an elevated mean foot temperature was associated with an increased risk for neuropathic foot ulceration. Diabetic patients with normal or low mean foot temperature were at risk for ischemic foot disease.

Another approach is to compare temperature maps of the individual's contralateral (left foot to their right) sites (an asymmetry analysis). This method has the advantage of being specific to the patient. However, it is dependent on geometrical symmetry between feet.[95] Therefore, preventive care is recommended when a patient is observed with temperature asymmetry exceeding 2.2°C (4°F) for at least two consecutive days between contralateral sites.[96] Using a remote temperature monitoring mat and the 2.2°C asymmetry approach, Frykberg et al.[97] predicted 97% of all non-acute plantar DFU on average of 35 days before clinical presentation with a specificity of 43%.

Nagase et al.[98] identified 20 patterns of thermal distribution to aid in diabetic foot assessment and surgical procedures. Bharara et al.[99] proposed a wound inflammatory index or temperature index for diabetic foot assessment, which is based on the difference in mean foot temperature and the wound bed, the area of the wound bed, and the area of the isotherm (highest or lowest temperature area).

Lavery et al.[100] in a multicenter study on 129 patients in remission, found that unilateral once-daily foot temperature monitoring can predict 91% of impending non-acute plantar foot ulcers on average of 41 days before clinical presentation.

In Figure 5.8, one can see thermographic assessment of Charcot arthropathy, which is characterized by the sudden onset of pain and inflammation in the foot of uncontrolled diabetic individuals.

5.1.5.10 Infection Detection
Infection also produces a localized increase in temperature as part of the body's immune response, which is triggered by the presence of pathogens in the affected area. Similar to that observed during inflammation, it is mainly driven by increased vascular permeability and blood flow.[101] Erythema, localized heat increases, and swelling are cardinal signs of wound infection; however, they are challenging to assess clinically. A study that compared the rate of clinical vs. histopathology-confirmed wound infection demonstrated that the accuracy of clinical inspection by trained observers is less than 70%.[102] To mitigate these shortcomings, multiple clinical decision rules have been developed to diagnose wound infections clinically, including the NERDS and STONEES mnemonics. While these rules offer a more objective assessment of a wound's characteristics, their

72 ■ Optical Methods for Managing the Diabetic Foot

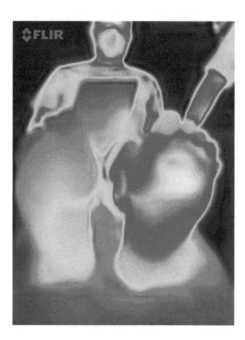

FIGURE 5.8 Thermographic assessment of Charcot arthropathy changes. Charcot arthropathy is characterized by the sudden onset of pain and inflammation in the foot of uncontrolled diabetic individuals. This disease is thermographically hallmarked by increased temperature in the complete foot, as opposed to a localized area in the case of infections. In this image, a significant increase in the temperature of the left foot can be observed. (Photograph courtesy of Dr. Jesus E Arriaga Caballero.)

sensitivities and specificities remain moderate, hovering at around 70% and 80%, respectively.[103] A study by Woo and Sibald[104] demonstrated that by combining any three clinical signs present in these rules, the specificity for infection detection rose to 90%, albeit with a reduction of the specificity to 70%. Interestingly, the same authors noted that the single-highest predictive sign of infection was a localized temperature increase, with an Odds Ratio (OR) of 8.05.

For this reason, thermography has risen as an adequate tool for helping to support the diagnosis of infection.[48] Rahbek et al. used this tool to discriminate between patients with orthopedic external pin infections and those without.[105] Their findings demonstrate that a cutoff value of 34°C for infection has a sensitivity of 73%; specificity of 67%; positive predictive value of 10%; and negative predictive value of 98%. Moreover, they also found an intra-rater agreement for thermography of ICC 0.85 (0.77–0.92), demonstrating that thermography can be used very reproducible by different observers, thus, removing some of the variability and subjectivity of clinical inspection alone. However, a significant pitfall of their study was that they used absolute temperature values, which have been found to have significant variation due to ambient factors and patient characteristics.[48]

Similarly, a study by Chanmugan et al.[106] demonstrated a temperature gradient in patients with inflamed vs. infected wounds. Those patients whose wounds were inflamed but not deemed to be clinically infected had a thermal load of +1.5°C to 2.2°C, compared to those with infected wounds, which were in the range of +4°C to 5°C. Additionally, they were able to document how in response to antibiotic therapies, the local temperature of the wounds dropped to +0.8°C to 1.1°C gradients, which were consistent with normal wounds. Another example of the use of thermography to detect inflammation and infection was provided by Oe et al.[107] who, in addition to detecting inflammation, found that thermography may be able to predict osteomyelitis, a severe complication of the diabetic wound before visible signs of infection are shown.

In Figure 5.9, one can see thermographic assessment of a diabetic ulcer infection. Under thermal imaging, deep tissue infections appear as localized, intense 'hotspots.'

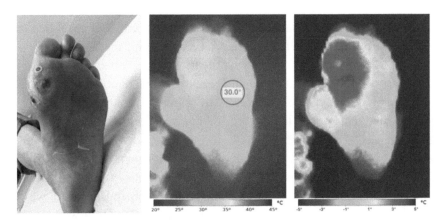

FIGURE 5.9 Thermographic assessment of a diabetic ulcer infection. Under thermal imaging, deep tissue infections appear as localized, intense 'hotspots,' usually above a thermal gradient of +2.5°C. In this case, the patient presented with an abscess around the first metatarsal area. Thermal imaging allowed its early identification and management.

5.1.5.11 Thermography: Practical Considerations

5.1.5.11.1 Core vs. Skin Temperature   Core body temperature refers to the temperature of the body's internal organs, such as the heart, liver, brain, and blood. The core body temperature is constant over time. The average normal body temperature is generally accepted as 37°C.

Unlike the core body temperature, which is kept constant, the skin temperature is subjected to changes and is usually lower than the core.

In realistic conditions, the skin temperature is the balance between heat production and radiative, convective, and evaporative heat losses. In,[108] it was found that other than the ambient temperature, blood perfusion and epidermis thickness are the primary factors responsible for skin temperature variations. In particular, the primary temperature drop in the skin is attributed to the cooling of the blood in the venous plexus. The temperature drop in the epidermis is on the scale of 0.1°C for the normal epidermis but can be 1.5–2°C or higher in calluses. Thus, local skin temperature variations can indicate epidermis thickness variations, particularly in callus-prone areas. Free moisture on the skin (e.g., wet wound) significantly increases the heat transfer, resulting in a temperature drop of several degrees Celsius. The relative air humidity significantly contributes (by slowing heat dissipation) only in the case of evaporative heat loss from wet skin. Therefore, wet skin is undesirable and should be avoided during a thermographic assessment.

5.1.5.11.2 Absolute vs. Relative Temperature   Interpreting an absolute skin temperature measurement (especially a single-point measurement) may pose a challenge because, as we just mentioned, it is affected by multiple factors, both internal and external. Therefore, a temperature gradient (the difference in temperatures between two points) is a more objective measure, which will depend less on the ambient conditions.

5.1.5.11.3 Wound Bed vs. Peri-wound   As has been mentioned already, the presence of free moisture on the skin (e.g., wet wound) significantly increases the heat transfer, resulting in a skin temperature drop, which can be on the scale of several degrees Celsius. It leads to a well-known problem of thermographic image distortion caused by evaporative water loss in the wound bed. This problem can be solved by allowing the wound to dry completely (which may delay the timing of the assessment) or by applying a non-permeable covering to the wound bed, eliminating the evaporation problem.[109]

5.1.5.11.4 Viewing Angle   The emissivity and apparent temperature of an object vary with the viewing angle. Thus, in practical applications, the viewing angle with respect to the tissue surface should not exceed 60°C or, ideally, 45°C, since for larger angles, the emissivity of the surface will be significantly reduced. Thus, for skin viewed obliquely, the lower emissivity may reduce the apparent temperature of more than 4°C, so a significant 'hotspot' in this region might be undetected in a thermogram.

5.1.5.11.5 Clinical Considerations   Several practical considerations should be considered for optimal thermal imaging.[110] Namely:

- Acclimation – it is recommended to allow some acclimation time before thermal imaging;
- Acute pain – capturing a thermal image immediately following a painful procedure is not advisable. The associated release of catecholamines with pain is associated with skin vasoconstriction;
- Smokers should be advised not to smoke for an hour before thermal imaging or to use tobacco or nicotine products to avoid vasoconstriction;
- Energy drinks containing high doses of caffeine and pseudoephedrine may also cause vasoconstriction;
- Volume depletion – it is not advisable to perform imaging if the patient is volume depleted. A common scenario is a patient with GI upset the day before with associated vomiting and/or diarrhea. If this is the case, it is better to allow 24 hours for volume repletion.
- Medications –antihypertensive medications or coffee in moderate doses will not affect thermal images.

5.1.5.12 Summary

IRT has significant advantages for screening evaluations of diabetic foot. Its low cost compared to other diagnostic techniques, ease of use, availability at the bedside, and reliability and repeatability collectively make IRT a 'practical' tool to determine if more costly evaluations are needed.

A screen for PAD thermography appears to be equal to ABI measurements, and in some cases, it is superior to ABI. It can be used in patients with noncompressible arterial calcinosis and is not subject to 'false negatives' stemming from the lower leg's trifurcated 'parallel' arterial supply. Thermography is useful for identifying the blood supply to free flaps and provides valuable data for postoperative flap surveillance. Similarly, it can assess the effectiveness of revascularization of the lower extremity and provide practical bedside surveillance of compromise.

Future work should focus on developing uniform methods for point measurements and asymmetry calculations. In some scenarios, Dynamic Thermography using provocative techniques, such as cooling, increases the signal-to-noise ratio and transforms subtle differences into easily interpreted results.

5.1.6 Exogenous Fluorescence Imaging

Several fluorescence-based technologies are used in wound care. Exogenous fluorescence imaging refers to fluorescence-based technologies which use an external fluorescent agent.

Fluorescence Angiography (FA) or Indocyanine Green Angiography (ICGA) is a novel optical modality that visualizes blood flow in vessels and perfusion. It uses an injectable dye, Indocyanine Green (ICG), injected into the systemic circulation and cleared exclusively via the liver. ICG has a documented record of safe clinical use. Another important feature of ICG is that it does not leak into extravascular space, allowing for multiple images in the same patient settings.[111] The technology was briefly described in Chapter 3 (see Section 3.3.3).

ICGA is particularly helpful for patients with Peripheral Arterial Disease (PAD). In,[112] the authors analyzed PAD patients with isolated infrapopliteal lesions who underwent ICGA. They compared the findings with the Ankle-Brachial Index (ABI) and Toe-Brachial Index (TBI) in 14 PAD patients and 9 control patients. The authors found that the Td 90% (the time elapsed from the maximum intensity to 90% of the maximum intensity) correlated most significantly with the ABI value. Thus, the authors concluded that ICGA might be helpful in quantitatively assessing peripheral perfusion.

Fluorescence Angiography can be used in DFU management. However, this type of imaging is typically confined to vascular labs and OR. While this method is highly validated and directly measures perfusion, it is invasive, poses the risk of anaphylactic reactions to the patient, and requires special equipment. Therefore, it is not routinely used in practice. The interested reader can look at specialized reviews[113] for further details.

In Figure 5.10, one can see Indocyanine Green imaging of a pressure wound.

5.1.7 Phosphorescent Long-Term Oxygen Sensors

Profusa Inc. (TX, USA) developed a method for tracking tissue oxygen in real-time with injectable, tissue-integrating microsensors.[114] The system consists of a small (500 μm × 500 μm × 5 mm), soft, flexible, tissue-like sensor from biocompatible hydrogels and a hand-held or wearable Bluetooth optical reader for intermittent or continuous noninvasive monitoring. The sensors are based on oxygen-dependent quenching of palladium porphyrin phosphorescence and are engineered to function for months to years in the body. In particular, dyes are synthetically engineered to emit in the Near-Infrared (optical window of the skin), covalently linked to the hydrogel backbone, and be stable under *in vivo* conditions. The system was able to monitor tissue oxygen for 9 months in a Sinclair mini-pig model.[115]

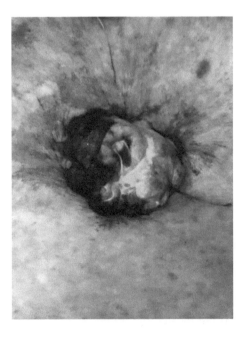

FIGURE 5.10 Indocyanine Green imaging of a pressure wound. Indocyanine Green was IV administered to assess the perfusion to a pressure injury wound bed. The image shows that nonperfused areas appear black, while those with adequate perfusion exhibit an intense fluorescence signal. (Photograph courtesy of Dr. Mario A. Martinez Jimenez.)

The method can be particularly useful for the early detection of restenosis after revascularization procedures.

5.2 INFECTION-TARGETING TECHNIQUES

5.2.1 Bacterial Fluorescence Imaging

Unlike exogenous fluorescence techniques, which use external agents to incite fluorescence, endogenous fluorescence techniques are based on tissue autofluorescence. While in tissue, there are many tissue molecules exhibiting autofluorescence, the two primary approaches relevant to wound care are assessment of metabolic (redox) tissue state (see Section 5.3.6) and bacterial visualization.

Bacterial fluorescence imaging is an endogenous fluorescence imaging technique based on the autofluorescence of bacterial metabolites. In particular, most clinically relevant bacteria produce porphyrins[116] (*S. aureus, S. epidermidis, Candida, S. marcescens, Viridians streptococci, Corynebacterium diphtheriae, S. pyogenes, Enterobacter,* and *Enterococcus*) or pyoverdine[117] (*P. aeruginosa*), which fluoresce (red and bluish-green fluorescence, respectively) when excited by incident light with a wavelength of 405 nm.

As a less-invasive and labor-intensive fluorescence imaging counterpart, bacterial fluorescence imaging is getting adopted in wound care, including point-of-care settings. The utility of fluorescence imaging for visualization of bacterial load in wounds has been demonstrated in numerous clinical studies, including studies with DFUs *in vivo*[118, 119, 120, 121] and *in vitro*.[122]

Fluorescence imaging may be used to distinguish between particular strains in the wound (e.g., *P. aeruginosa*), to assess (qualitatively or semi-quantitatively) bacteria present in the wound (infection detection), or to guide sampling, debridement, antimicrobial selection, grafting, and the use of Cell Tissue Products (CTP). An example of bacterial autofluorescence is depicted in Figure 5.11.

5.2.1.1 Infection Detection

The fluorescence imaging demonstrated its utility in identifying bacterial loads more than 10^4 CFU/g close to real-time. In particular, Farhan et al.[44] identified 11 clinical studies (including 613 wounds with various etiology), which aimed to assess the diagnostic accuracy for detecting bacterial loads of $\geq 10^4$ CFU/g and calculated weighted averages for sensitivity (74%), specificity (88%), positive predictive value (91%), negative predictive

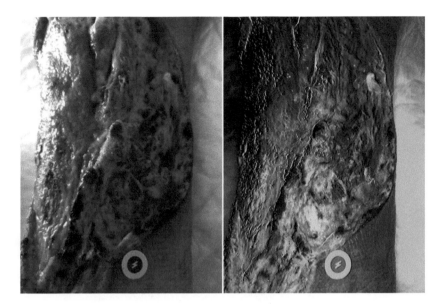

FIGURE 5.11 Bacterial autofluorescence. Bacterial pigments are easily visualized in wound beds after excitation with a 405 nm light source. Species producing heme-containing compounds fluoresce in red, and those producing pyoverdine in cyan. This photograph shows polybacterial contamination, as both cyan and red fluorescence are captured.

value (53%), and overall diagnostic accuracy (75%). It represents a threefold to fourfold increase in sensitivity compared to CSS alone.

5.2.1.2 Sampling and Debridement Guidance

Another practical application of fluorescent imaging is to guide debridement and sampling (swabbing or biopsy). A pilot evaluation compared standard Levine swab results with fluorescence-guided curettage samples and found that the Levine technique gave a 36% false-negative laboratory report.[123] An example of using bacterial fluorescence imaging for debridement guidance is depicted in Figure 5.12.

5.2.1.3 Treatment Selection

The ultimate utility of fluorescence imaging is to assist in treatment selection. Studies have shown that fluorescence imaging can prompt changes in proposed treatment plans, including alterations in antimicrobial prescribing,[124] decisions around negative pressure wound therapy,[125] and timing of grafting or applications of Cell Tissue Products (CTP).[126] Bacterial

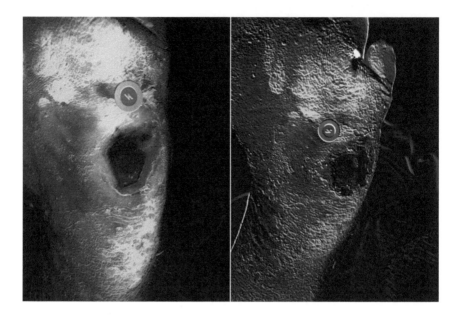

FIGURE 5.12 Targeted debridement using bacterial fluorescence. Bacterial fluorescence can be used to identify areas where excessive bacterial overgrowth is present, as the lower limit of detection of this technology is 10^4 CFU/g organisms. In these images, the left panel shows a wound pre-debridement and the right one shows the same wound post-debridement. Bright areas of red fluorescence indicate bacteria closer to the wound's surface, while pink areas represent subsurface bacteria. Noteworthy is that keratin exhibits autofluorescence in the white–cyan range. Care must be taken so this signal is not interpreted as bacterial contamination due to pseudomonas.

fluorescence imaging leads to changes in the treatment plan in as many as 73% of cases.

The debut of new optical technology raises practical questions of when and where it should be used in daily practice. To better understand the role of fluorescence imaging, a Delphi consensus method[127] was used to determine fluorescence imaging guidelines for chronic ulcers.[128] Surveying a panel of experts using fluorescence in multiple care settings found >80% reported changes in treatment plans, 96% reported that imaging-informed treatment plans led to improved wound healing, 83% reported reduced rates of microbiological sampling, and 78% reported reduced rates of amputations.

There is a universal agreement for eliminating or reducing bioburden as a prerequisite for healing. However, the sensitivity of clinical examination is less than 15%, and more than 80% of wounds with high bacterial loads remain undetected.[129]

Chronic wounds are more common in patients with diabetes, the elderly, and patients on high-dose steroids, all of whom impaired immune function and frequently fail to mount Clinical Signs and Symptoms (CSS) of infection.

The Delphi method revealed a high agreement with the sequence of steps outlined in the clinical workflows for wound cleansing, debridement, and wound-bed assessment for readiness to receive grafts or Cell Tissue Product therapies (see Figure 5.13).

If these fluorescence-imaging guidelines are followed, does it really change the healing rate? To answer that, a Randomized Controlled Trial (RCT) was designed to compare healing rates and the decision-making associated with fluorescence imaging.[130]

The primary outcome was the proportion of ulcers healed at 12 weeks by blinded assessment. Secondary outcomes included wound area reduction

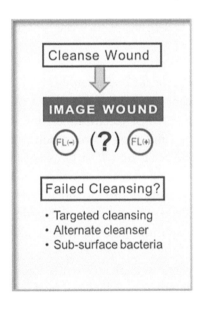 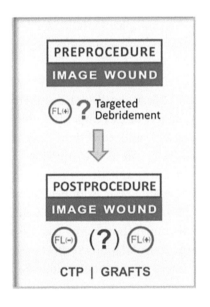

FIGURE 5.13 Fluorescence Imaging Guidelines for Wound Bacterial Burden Based on Delphi Consensus. Two common scenarios are illustrated. The first is a cleansing scenario (left). Fluorescence imaging provides guidance for the adequacy of cleansing. Scenario 2 (right panel) provides procedural guidance for wound debridement and guidance to determine if the wound bed is 'clean' enough to receive advanced Cell Tissue Product (CPT) therapy or skin grafts. (Reproduced from[113] under CC BY 4.0 license.)

at 4 and 12 weeks and change in management decisions after fluorescence imaging.

The proportion of ulcers healed at 12 weeks in the autofluorescence arm was 45% vs. 22% in the control arm. Wound area reduction was 40.4% (fluorescence positive) vs.38.6% (control) at 4 weeks and 91.3% (fluorescence positive) vs. 72.8% (control) at 12 weeks.

Wound debridement was the most common intervention in wounds with positive fluorescence imaging. There was a stepwise trend in healing favoring those with negative autofluorescence imaging, followed by those with positive autofluorescence who had an intervention. The poorest outcome was the group with positive fluorescence and no corrective intervention.

Time to heal is important. Slow healing is associated with infection, hospitalization, and amputation. DFUs present for >30 days have fivefold increased odds of infection compared with those that heal, and infection increases the odds of hospitalization 55-fold and odds of amputation 154-fold when compared with noninfected DFUs.[131]

An example of a failed CPT graft due to significant bacterial contamination is shown in Figure 5.14.

5.2.1.4 Bacterial Fluorescence Imaging: Practical Consideration

The ultimate utility of bacterial fluorescence imaging is treatment plan guidance. An example is differentiating between colonization (no host reaction) and infection (host reaction). Confusing colonization with infection can lead to spurious associations that may lead to expensive, ineffective, and time-consuming interventions. Some scenarios can be challenging. For example, immunocompromised patients do not show signs of infection as normal patients. Neutropenic patients (<500 neutrophils/mm^3) show no pyuria, no purulent sputum, little infiltrate, and no large consolidation on chest X-ray.

There is a temptation to use a single metric to characterize the stage of bacterial presence. Such metric can be bioburden or bacterial load measured in colony-forming units or CFU per gram [CFU/g]. In particular, researchers have shown that when bioburden reaches levels >10^5 CFU/g of tissue, it can cause wound infection.[132,133] When bioburden levels >10^6 CFU/g, wound healing is impeded,[133] and the likelihood of a successful skin graft approaches zero. In another study on eight DFU patients, authors found that wounds with a bioburden >10^6 CFU/g had a healing rate of 0.05 cm per week.[134] When bioburden ranged between 10^5 and 10^6

Optical Diagnostic Techniques ■ 83

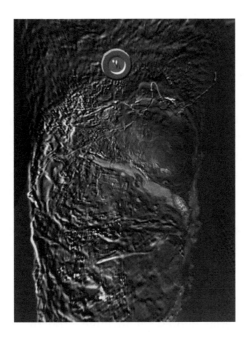

FIGURE 5.14   Failed CPT graft due to bacterial contamination. A failed CPT graft due to significant bacterial contamination is shown here. Bacterial fluorescence imaging can be used to assess the wound bed before grafting and to monitor the graft's evolution before significant contamination threatens its survival.

CFU/g, the wound healing rate was 0.15 cm per week. When patients had no bioburden in the wound, the healing rate was fastest at 0.2 cm of epithelialization per week.[135]

Using bacterial fluorescence imaging for qualitative assessment faces several challenges. Firstly, the boundary between colonization and infection is subject to various definitions. Different authors introduce various definitions, ranging from 10^4 CFU/g to 10^5 CFU/g. Researchers have proposed that there is a critical point between colonization and infection, termed critical colonization.[134] However, it is more of a theoretical concept at this point, which has yet to be established. One must also consider the specific 'type' of bacteria, as some strains are more virulent than others. For example, Group A *Streptococcus pyogenes* is more virulent than *Streptococcus viridans*. In other words, 10^3 CFU/g of *S. pyogenes* is more severe than 10^4 CFU/g of *S. viridans*.

Secondly, it is difficult to establish the direct correlation between the fluorescence intensity and the bacterial load (bioburden). As both pyoverdine and porphyrins are by-products of bacterial presence, there is no simple relationship between their concentration and bioburden, as numerous factors can impact their production and fluorescence. For example, pyoverdine fluorescence is quenched in the presence of Fe^{2+} and Fe^{3+}.[135]

While using bacterial fluorescence imaging, good imaging practices are necessary, such as cleaning the wound, removing as much blood as possible, and removing imaging artifacts from white bedsheets and gauze bandages when possible. Blood, in particular, can absorb the violet excitation light and mask other fluorescence signatures if not removed.

In addition, it should be considered that the violet excitation light cannot penetrate more than 1 mm to 1.5 mm into the skin. While this enables the detection of some subsurface bacteria, the presence of bacteria deeper within the wound tissue may not be visible, including deep tunneling infection.

5.2.1.5 Bacterial Fluorescence: Image Interpretation

To interpret bacterial fluorescent imaging properly, the user needs to understand anatomic sources of autofluorescence. The color of the fluorescent emission is a function of the Stokes shift (see Appendix B, Bio Optics Primer). Connective tissue (collagen, elastin, fibrin), melanin, tendon, and bone are all fluorescent, with light emissions spanning a broad range of colors in the visible spectrum, between 420 nm and 700 nm. Some of these wavelengths can be removed from the image using custom emission filters (e.g., bandpass filters). For example, only fluorescence with wavelengths between 501 nm and 542 nm (cyan and green) and wavelengths between 601 nm and 664 nm (orange and red) can pass through to the camera. The emission filter also removes non-absorbed (reflected) excitation light from the image.

The fluorescent image can also be impacted by a skin tone (see Figure 5.15).

The dominant color of bacterial fluorescence images depends on a particular implementation of fluorescence imaging technology. For example, when viewing a wound using a Moleculight imager (Moleculight, Toronto, Canada), the dominant color is within the green spectrum (500–570 nm), which can be attributed to collagen and elastin. The shade will vary with tissue type and the associated densities of collagen, elastin, and their degree of cross-linking. Tendon and bone have the highest collagen content and fluoresce a very bright green to glowing white color. Characteristic fluorescence signals from wound tissues are depicted in Figure 5.16.

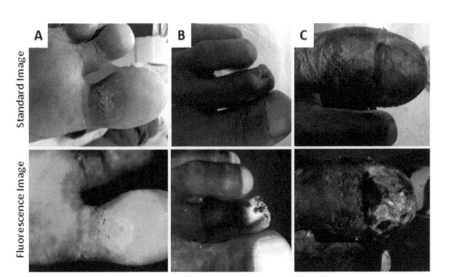

FIGURE 5.15 Variation in green fluorescence due to skin tone. (A) Lighter skin tones fluoresce a light green color. As melanin concentration increases (B, C), the tone of green in the fluorescence images also darkens due to the absorption of excitation light preferentially by melanin causing attenuation of the tissue fluorescence. (Reproduced from under CC BY 4.0 license.) MY Rennie, D Dunham, L Lindvere-Teene, et al. "Understanding real-time fluorescence signals from bacteria and wound tissues observed with the MolecuLight i:XTM", *Diagnostics (Basel)*, 9(1):22 (2019).

Red fluorescence is associated with bacteria and is a function of bacterial depth within the tissue and other fluorophores that contribute to the image's final 'composite' color. Imaging depth is limited by the penetration of the source light (405 nm), which is a maximum of 1.5 mm. The brightest red colors are from fluorescent bacteria at or close to the surface, while pink and faint red are subsurface bacteria. The color change is a function of numerous factors, including loss of the excited light with depth, scattering of the fluorescent light emerging from the tissue, and color 'blending' with overlying fluorophores. As such, the bacterial fluorescence signals (red or cyan) can vary based on the depth of the bacteria and the 'blending' of bacterial emission wavelengths with wavelengths from nonbacterial sources, e.g., blood, pus, and wound fluid.

Figure 5.17 provides some examples of bacterial fluorescence.

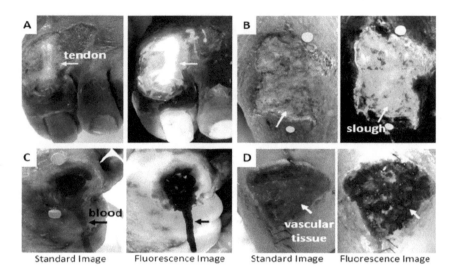

FIGURE 5.16 Characteristic fluorescence signals from wound tissues. (A) Intense green/glowing white fluorescence from a tendon due to high collagen content. This signal could be confused for *Pseudomonas aeruginosa* if the clinician did not compare the fluorescence observed with the standard image of the wound. The standard image shows that the glowing white region of fluorescence is a tendon. Bright green is also shown in a toenail. (B) Slough, which appears yellow on a standard image, fluoresces light green in this ulcer due to its high fibrin content. (C) Blood from a DFU (post-debridement) appears maroon on fluorescence images due to hemoglobin's absorption of the excitation light. Blood should always be removed from the wound bed before imaging. Blush red fluorescence can be seen in the peri-wound tissue of this DFU, suggesting the presence of bacteria at moderate to heavy loads. (D) Highly vascular tissue (e.g., granulation tissue) also fluoresces maroon due to hemoglobin's absorption of illumination light. (Reproduced from[121] under CC BY 4.0 license.)

Image artifacts can occur from white light contamination and nonbiologic sources such as tattoos, cotton linens, and selected wound-cleansing solutions. Figure 5.18 illustrates some confounding phenomena that should be considered to interpret bacterial fluorescence properly.

5.3 OTHER TECHNIQUES

In this section, we list several relatively mature techniques, which are used (or can be used) in DFU research, but due to various reasons, are further from translation to PoC modalities than the ones mentioned above.

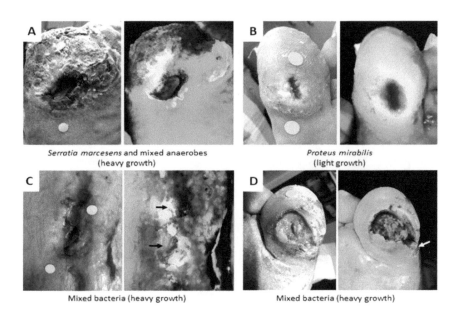

FIGURE 5.17 Fluorescence from bacteria appears red or pink/blush. Red fluorescence is produced by endogenous porphyrins (produced by most bacterial species) when excited by 405 nm violet light. All wounds show a standard image on the left and a fluorescence image on the right, with microbiology results from cultures reported below. Circular yellow stickers are present. (Reproduced from[121] under CC BY 4.0 license.)

5.3.1 Orthogonal Polarization Spectral Imaging

Orthogonal Polarization Spectral (OPS) imaging is based on illuminating skin tissue with linearly polarized light and collecting resultant depolarized photons scattered by the tissue components using a polarizer positioned orthogonal to the plane of illuminating light.[136] It increases sensitivity by discarding single scattered photons, mostly from specular (Fresnel) reflection on surfaces. In OPS imaging, the microvasculature is illuminated with polarized green light. The technique can be used to visualize vessels and Red Blood Cells (which absorb light in the green range of the spectrum). Thus, it can be particularly helpful in dermatology and rheumatology. In wound care, the method has been mainly applied to detect burn depth.[137] However, it was also used on wounds with other etiologies (e.g., chronic venous insufficiency).[138]

88 ■ Optical Methods for Managing the Diabetic Foot

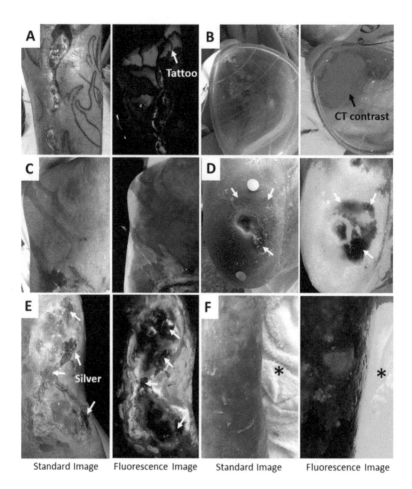

FIGURE 5.18 Confounding signals. (A) Bright red fluorescence from a tattoo. (B) A fluorescent red CT contrast agent was observed within a Jackson Pratt wound drainage system for a patient undergoing a sinogram. (C) Pink chlorohexidine on a patient's foot and ankle. This solution is tinted with pink to let clinicians see where the tissue has been cleaned; the tinting appears red/pink on fluorescence images. (D) Images of a wound after application of iodine during cleaning. Arrows denote the region of iodine, which appears reddish brown on the standard image and dark on the fluorescence image. (E) Venous leg ulcer with the residual silver product (white arrows) on the wound. Regions with silver products on standard images result in regions of black on fluorescence images, as the violet illumination light cannot penetrate through the silver. Silver products should be removed from the wound before imaging, if possible. (F) Bright fluorescence from a white bed sheet (asterisk) in the field of view. Clinicians should be aware that white items in the field of view may fluoresce. (Reproduced from[121] under CC BY 4.0 license.)

5.3.2 Other Surface Reflection Discarding Techniques

Similar to Orthogonal Polarization Spectral imaging, several methods (typically contact modalities) exploit the idea of discarding surface reflectance. Dark Field microscopy is one such modality. The Sidestream Dark Field (SDF) imaging device discards the surface-scattered photons by optically isolating the light guide from the illuminating outer ring. The SDF device comprises a central light guide in contact with the skin, surrounded by concentrically placed light-emitting diodes emitting green light. Goedhart et al.[139] developed and validated the SDF technology against OPS on nailfold capillaries.

The technology subsequently evolved into Incident Dark Field (IDF) technique. The IDF is based on the method initially developed by Sherman et al.[140] Gilbert-Kawai et al.[141] compared IDF with SDF in sublingual microcirculatory imaging and found it superior.

In addition to Dark Field microscopy, several other contact imaging techniques have been proposed: lensless microscopy,[142] which uses multicore fibers, and numerical aperture-gated microscopy.[143]

5.3.3 Confocal Microscopy

Confocal microscopy or Confocal Laser Scanning Microscopy (CLSM) is a high-resolution optical detection technique that allows monitoring specimens in multiple planes of depth by adjusting a specific focal depth. While CLSM has been used to image cutaneous wounds during healing noninvasively, it is currently limited to superficial epidermal wounds.[144] In addition, it has been used to distinguish between normal and abnormal skin morphology.[145] The closely related reflectance mode confocal microscopy allows for the simultaneous recording and evaluation of microcirculation, histomorphology, and inflammatory cell trafficking. It has been used successfully for the study of burn wounds.[146]

However, it should be noted that the techniques mentioned above are microscopic and limited to areas with thin epidermis or mucosa. Therefore, their applications to wound care in general, and diabetic foot in particular, can be quite limited.

5.3.4 Speckle Plethysmography

Ghijsen et al.[147] proposed Speckle Plethysmography (SPG), a wearable technology based on speckle imaging for characterizing microvascular flow and resistance. Razavi et al.[148] compared SPG with Ankle-Brachial Index (ABI), Toe-Brachial Index (TBI), and clinical presentation of patients per

Rutherford category on 167 limbs (90 patients). Qualitatively, they found that SPG is analogous to Doppler velocity measurements, and waveform phasicity and amplitude degradation were observed with increasing PAD severity.

5.3.5 Perfusion Imaging

Several groups recently proposed different versions of perfusion imaging. This group of technologies is based on remote Photoplethysmography (rPPG), also called imaging Photoplethysmography (iPPG)[149] or video photoplethysmography. It creates a 2D map of the amplitude of blood pulsations in a target area. The technology is based on the observations that the blood pulsations caused by pulse propagation vary in amplitude depending on multiple physiological parameters, including local hydrostatic pressure. The technology can identify low-perfused areas (e.g., occlusions)[150] and monitor reperfusion intra-operatively. In particular, it was used for reperfusion monitoring during abdominal surgery [151] and free microvascular anastomosed fasciocutaneous flaps.[151]

5.3.6 Metabolic Fluorescence Imaging

Metabolic fluorescence imaging or Optical Metabolic Imaging (OMI) is a noninvasive, high-resolution, quantitative tool for monitoring cellular metabolism. It refers to a group of technologies that rely on measurements of the fluorescence intensity of NAD(P)H and FAD.

They are based on the observation that the ratio of these fluorophores (NADH/FAD), called redox ratio, acts as a quantitative marker of the mitochondrial redox state of the tissue.[152] Here, redox (reduction–oxidation) refers to a type of chemical reaction in which the oxidation states of the substrate change.

In addition to the original redox scanner developed in the 1970s by Britton Chance,[153] several other technologies have been developed since then, including epifluorescence microscopy, confocal microscopy,[154] Two-Photon Excitation Fluorescence (TPEF) microscopy,[155] and fluorescence lifetime microscopy (FLIM).[156] As such, they enabled a noninvasive 3-D visualization of cell and tissue redox states. They have numerous applications in the metabolic monitoring of cells and tissues in cancer, neuroscience, and tissue engineering. Gradually, it is getting applied to DFU research as well. For example, in,[157] the technology was used to quantify the differences between the redox state of wounds in diabetic mice and the control mice. A significant correlation was observed between the redox state and the area of the wounds.

However, despite significant advances in recent years, these technologies are microscopy-based and still confined to research labs. For the current state of this technology, readers can refer to recent reviews (see, e.g.).[158]

NOTES

1. J Kluz, R Malecki, R Adamiec, "Practical importance and modern methods of the evaluation of skin microcirculation during chronic lower limb ischemia in patients with peripheral arterial occlusive disease and/or diabetes", *Int Angiol*, 32:42–51 (2013).
2. O Bongard, B Fagrell, "Discrepancies between total and nutritional skin microcirculation in patients with peripheral arterial occlusive disease (PAOD)", *Vasa*, 19:105–111 (1990).
3. G Jörneskog, K Brismar, B Fagrell, "Skin capillary circulation is more impaired in the toes of diabetic than non-diabetic patients with peripheral vascular disease", *Diabet Med*, 12:36–41 (1995).
4. G Jörneskog, "Why critical limb ischemia criteria are not applicable to diabetic foot and what the consequences are", *Scand J Surg*, 101:114–118 (2012).
5. LX Zhan, BC Branco, DG Armstrong, et al. "The society for vascular surgery lower extremity threatened limb classification system based on Wound, Ischemia, and foot Infection (WIfI) correlates with risk of major amputation and time to wound healing", *J Vasc Surg*,61(4):939–944 (2015).
6. E Raposio, N Bertozzi, R Moretti, et al. "Laser doppler flowmetry and transcutaneous oximetry in chronic skin ulcers: A comparative evaluation", *Wounds*, 29(7):190–195 (2017).
7. ME Gschwandtner, E Ambrózy, B Schneider, et al. "Laser doppler imaging and capillary microscopy in ischemic ulcers", *Atherosclerosis*, 3(142):225–232 (1999).
8. N Morimoto, N Kakudo, PV Notodihardjo, et al. "Comparison of neovascularization in dermal substitutes seeded with autologous fibroblasts or impregnated with bFGF applied to diabetic foot ulcers using laser Doppler imaging", *J Artif Organs*,16(17):352–357 (2014).
9. AK Murray, AL Herrick, TA King, "Laser Doppler imaging: A developing technique for application in the rheumatic diseases", *Rheumatology*, 43(10): 1210–1218 (2004).
10. A Bircher, EM de Boer, T Agner, et al. "Guidelines for measurement of cutaneous blood flow by laser Doppler flowmetry. A report from the Standardization Group of the European Society of Contact Dermatitis", *Contact Dermatitis*, 30:65–72 (1994).
11. JM Gardner-Medwin, IA Macdonald, JY Taylor, et al. "Seasonal differences in finger skin temperature and microvascular blood flow in healthy men and women are exaggerated in women with primary Raynaud's phenomenon", *Br J Clin Pharmacol*, 52:17–23 (2001).

12. TMAJ van Vuuren, C van Zandvoort, S Doganci, et al. "Prediction of venous wound healing with laser speckle imaging", *Phlebology*, 32:658–664 (2017).
13. M Hellmann, M Roustit, F Gaillard-Bigot, JL Cracowski, "Cutaneous iontophoresis of treprostinil, a prostacyclin analog, increases microvascular blood flux in diabetic malleolus area", *Eur J Pharmacol*, 758:123–128 (2015).
14. M Neidrauer, L Zubkov, MS Weingarten, et al. "Near infrared wound monitor helps clinical assessment of diabetic foot ulcers", *J Diabetes Sci Technol*, 4(4):792–798 (2010).
15. S Anand, N. Sujatha, VB Narayanamurthy, et al. "Diffuse reflectance spectroscopy for monitoring diabetic foot ulcer – A pilot study", *Optics and Lasers in Eng*, 53: 1–5 (2014).
16. SM Rajbhandari, ND Harris, S Tesfaye, JD Ward, "Early identification of diabetic foot ulcers that may require intervention using the micro light guide spectrophotometer", *Diabetes Care*, 22(8):1292–1295 (1999).
17. S Poosapadi Arjunan, AN Tint, B Aliahmad, et al. "High-resolution spectral analysis accurately identifies the bacterial signature in infected chronic foot ulcers in people with diabetes", *Int J Low Extrem Wounds*, 17(2):78–86 (2018).
18. G Lu, B Fei, "Medical hyperspectral imaging: A review", *J Biomed Opt*, 19: 010901 (2014).
19. L Khaodhiar, T Dinh, KT Schomacker, et al. "The use of medical hyperspectral technology to evaluate microcirculatory chaves in diabetic foot ulcers and to predict clinical outcomes", *Diabetes Care*, 30(4): 903–910 (2007).
20. A Nouvong, B Hoogwerf, E Mohler, et al. "Evaluation of diabetic foot ulcer healing with hyperspectral imaging of oxyhemoglobin and deoxyhemoglobin", *Diabetes Care*, 32(11):2056–2061 (2009).
21. G Saiko, P Lombardi, Y Au, et al. "Hyperspectral imaging in wound care: A systematic review", *Int Wound J*, 17(6): 1840–1856 (2020).
22. RL Greenman, S Panasyuk, X Wang, et al. "Early changes in the skin microcirculation and muscle metabolism of the diabetic foot", *Lancet*, 366(9498):1711–1717 (2005).
23. TE Serena, R Yaakov, L Serena, et al. "Comparing near infrared spectroscopy and transcutaneous oxygen measurement in hard-to-heal wounds: A pilot study", *J Wound Care*, 29(Sup6):S4–S9 (2020).
24. WJ Jeffcoate, DJ Clark, N Savic, et al. "Use of HSI to measure oxygen saturation in the lower limb and its correlation with healing of foot ulcers in diabetes", *Diabet Med*, 32(6):798–802 (2015).
25. A Landsman, "Visualization of wound healing progression with near infrared spectroscopy: A retrospective study", *Wounds*, 32(10):265–271 (2020).
26. D Yudovsky, A Nouvong, K Schomacker, L Pilon, "Assessing diabetic foot ulcer development risk with hyperspectral tissue oximetry", *J Biomed Opt*, 16(2):026009 (2011).
27. D Yudovsky, L Pilon, "Rapid and accurate estimation of blood saturation, melanin content, and epidermis thickness from spectral diffuse reflectance", *Appl Opt*, 49(10):1707–1719 (2010).

28. HN Mayrovitz, A McClymont, N Pandya, "Skin tissue water assessed via tissue dielectric constant measurements in persons with and without diabetes mellitus", *Diab Tech & Theur*, 15(1): 1–6 (2013).
29. M Attas, T Posthumus, B Schattka, et al. "Long-wavelength near-infrared spectroscopic imaging for in-vivo skin hydration measurements", *Vibrational Spectroscopy*, 28(1): 37–43 (2002).
30. G Saiko, "On the feasibility of skin water content imaging adjuvant to tissue oximetry", *Adv Exp Med Biol*, 1269: 191–195 (2021).
31. G Saiko, "Callus thickness determination adjuvant to tissue oximetry imaging", In Proceedings of the 15th International Joint Conference on Biomedical Engineering Systems and Technologies (BIOSTEC 2022) – Volume 2: BIOIMAGING, 147–152.
32. K Kounas, T Dinh, K Riemer, et al. "Use of hyperspectral imaging to predict healing of diabetic foot ulceration", *Wound Rep Reg*, 31(2):199–204 (2023).
33. L Khaodhiar, T Dinh, KT Schomacker, et al. "The use of medical hyperspectral technology to evaluate microcirculatory changes in diabetic foot ulcers and to predict clinical outcomes", *Diabetes Care*, 30:903–910 (2007).
34. A Nouvong, B Hoogwerf, E Mohler, et al. "Evaluation of diabetic foot ulcer healing with hyperspectral imaging of oxyhemoglobin and deoxyhemoglobin", *Diabetes Care*,32:2056–2061 (2009).
35. WJ Jeffcoate, DJ Clark, N Savic, et al. "Use of HSI to measure oxygen saturation in the lower limb and its correlation with healing of foot ulcers in diabetes", *Diabet Med*,32:798–802 (2015).
36. DJ Cuccia, FP Bevilacqua, AJ Durkin, et al. "Quantitation and mapping of tissue optical properties using modulated imaging", *J Biomed Opt*, 14(2): 024012 (2009).
37. V Lukic, VA Markel, JC Schotland, "Optical tomography with structured illumination", *Opt Lett*, 34(7):983–985 (2009).
38. S Konecky, A Mazhar, D Cuccia, et al. "Quantitative optical tomography of sub-surface heterogeneities using spatially modulated structured light", *Opt Express*, 17: 14780–14790 (2009).
39. S Gioux, A Mazhar, DJ Cuccia, "Spatial frequency domain imaging in 2019: Principles, applications, and perspectives", *J Biomed Opt*, 24(7): 071613 (2019).
40. M Saidian, JRT Lakey, A Ponticorvo, et al. "Characterisation of impaired wound healing in a preclinical model of induced diabetes using wide-field imaging and conventional immunohistochemistry assays", *Int Wound J*, 16(1): 144–152 (2018).
41. GA Murphy, RP Singh-Moon, A Mazhar, et al. "Quantifying dermal microcirculatory changes of neuropathic and neuroischemic diabetic foot ulcers using spatial frequency domain imaging: A shade of things to come?" *BMJ Open Diabetes Res Care*, 8(2):(2020).
42. https://clinicaltrials.gov/ct2/show/NCT03341559 as was visited on May 12, 2021.

43. J Martín-Vaquero, A Hernández Encinas, A Queiruga-Dios, et al. "Review on wearables to monitor foot temperature in diabetic patients", *Sensors (Basel)*, 19(4):776 (2019).
44. RN Lawson, GD Wlodek, DR Webster, "Thermographic assessment of burns and frostbite", *Canad Med Assos J*, 84: 1129–1131 (1961).
45. M Bharara, JS Choess, DG Armstrong, "Coming events cast their shadows before: Detecting inflammation in the acute diabetic foot and the foot in remission", *Diabetes Metab Res Rev*, 28 Suppl 1: 15–20 (2012).
46. JL Ramirez-GarciaLuna, R Bartlett, JE Arriaga-Caballero, et al. "Infrared thermography in wound care, surgery, and sports medicine: A review", *Front Physiol*, 13:838528 (2022).
47. L De Weerd, JB Mercer, S Weum, "Dynamic infrared thermography", *Clin Plast Surg*, 38:277–292 (2011).
48. S Bagavathiappan, T Saravanan, J Philip, et al. "Investigation of peripheral vascular disorders using thermal imaging", *Br J Diabetes*, 8: 102–104 (2008).
49. J Philip, T Jayakumar, B Raj, et al. "Infrared thermal imaging for detection of peripheral vascular disorders", *J Med Phys*, 34: 43–47 (2009).
50. M Carabott, C Formosa, A Mizzi, et al. "Thermographic characteristics of the diabetic foot with peripheral arterial disease using the angiosome concept", *Exp Clin Endocrinol Diabetes*, 129: 93–98 (2021).
51. WC Chang, CY Wang, Y Cheng, et al. "Plantar thermography predicts freedom from major amputation after endovascular therapy in critical limb ischemic patients", *Medicine*, 99: e22391 (2020).
52. AJ de Carvalho Abreu, RA de Oliveira, AA Martin, "Correlation between ankle-brachial index and thermography measurements in patients with peripheral arterial disease", *Vascular*, 30: 88–96 (2022).
53. A Gatt, O Falzon, K Cassar, et al. "The application of medical thermography to discriminate neuroischemic toe ulceration in the diabetic foot", *Int J Low Extremity Wounds*, 17: 102–105 (2018).
54. A Gatt, K Cassar, O Falzon, et al. "The identification of higher forefoot temperatures associated with peripheral arterial disease in type 2 diabetes mellitus as detected by thermography", *Prim Care Diabetes*, 12: 312–318 (2018).
55. Y Hosaki, F Mitsunobu, K Ashida, et al. "Non-invasive study for peripheral circulation in patients with diabetes mellitus", *OKAYAMA Univ Sci Achiev Repos*, 72: 31–37 (2002).
56. CL Huang, YW Wu, CL Hwang, et al. "The application of infrared thermography in evaluation of patients at high risk for lower extremity peripheral arterial disease",*J Vasc Surg*,54(4):1074–1080 (2011).
57. A Ilo, P Romsi, J Mäkelä, "Infrared thermography and vascular disorders in diabetic feet", *J Diabetes Sci Technol*, 14(1):28–36 (2020).
58. A Ilo, P Romsi, M Pokela, J Mäkelä, "Infrared thermography follow-up after lower limb revascularization", *J Diabetes Sci Technol*, 15: 807–815 (2021).
59. CFJ Renero, A Ziga-Martínez, M Silva-González, V Carbajal-Robles, "The peripheral artery disease through the thermogram and the photoplethysmogram before and after a revascularization surgery", *J Diabetes Sci Technol*, 15: 1200–1201 (2021).

60. E Staffa, V Bernard, L Kubíček, et al. "Using Noncontact infrared thermography for long-term monitoring of foot temperatures in a patient with diabetes mellitus", *Ostomy Wound Manag*, 62: 54–61 (2016).
61. GA Wallace, N Singh, E Quiroga, NT Tran, "The use of smart phone thermal imaging for assessment of peripheral perfusion in vascular patients", *Ann Vasc Surg*, 47: 157–161 (2018).
62. H Wang, DR Jr Wade, J Kam, "IR imaging of blood circulation of patients with vascular disease", In *Thermosense XXVI*; SPIE, Bellingham, WA, USA, 2004, pp. 115–123.
63. G Zenunaj, N Lamberti, F Manfredini, et al. "Infrared thermography as a diagnostic tool for the assessment of patients with symptomatic peripheral arterial disease undergoing infrafemoral endovascular revascularisations", *Diagnostics*, 11: 1701 (2021).
64. G McQuilkin, D Panthagani, R Metcalfe, et al. "Digital thermal monitoring of vascular reactivity closely correlates with Doppler flow velocity", *Conf Proc IEEE Eng Med Biol Soc*, 2009:1100e3 (2009).
65. O Iida, M Takahara, Y Soga, et al. "SPINACH investigators. Three-year outcomes of surgical versus endovascular revascularization for critical limb ischemia: The SPINACH study (Surgical Reconstruction Versus Peripheral Intervention in Patients with Critical Limb Ischemia)", *Circ Cardiovasc Interv*, 10(12):e005531 (2017).
66. T Pakarinen, A Joutsen, N Oksala, A Vehkaoja, "Assessment of chronic limb threatening ischemia using thermal imaging", *J Therm Biol*,112:103467 (2023).
67. J Pasek, A Stanek, G Cieslar, "The role of physical activity in prevention and treatment of peripheral vascular disorders", *Acta Angiol*, 26: 19–27 (2020).
68. T Kawasaki, T Uemura, K Matsuo, et al. "The effect of different positions on lower limbs skin perfusion pressure", *Sep Indian J Plast Surg*, 46(3): 508–512 (2013).
69. L Janský, V Vavra, P Janský, et al. "Skin temperature changes in humans induced by local peripheral cooling", *J Therm Biol*, 28(5): 429–437 (2003).
70. M Kaczmarek, A Nowakowski, "Active IR-thermal imaging in medicine", *J Nondestr Eval*,35 (2016).
71. GG Hallock, "Dynamic infrared thermography and smartphone thermal imaging as an adjunct for preoperative, intraoperative, and postoperative perforator free flap monitoring", *Plast Aesthet Res*, 6:29 (2019).
72. JT Hardwicke, O Osmani, JM Skillman, "Detection of perforators using smartphone thermal imaging", *Plast Reconstr Surg*, 137:39–41 (2016).
73. N Pereira, "Reply: Detection of perforators for free flap planning using smartphone thermal imaging: A concordance study with computed tomographic angiography in 120 perforators", *Plast Reconstr Surg*, 142:605e (2018).
74. Y Itoh, K Arai, "Use of recovery-enhanced thermography to localize cutaneous perforators", *Ann Plast Surg*, 34:507–511 (1995).
75. L de Weerd, JB Mercer, LB Setsá, "Intraoperative dynamic infrared thermography and free-flap surgery", *Ann Plast Surg*, 57:279–284 (2006).

76. MV Muntean, S Strilciuc, F Ardelean, AV Georgescu, "Dynamic infrared mapping of cutaneous perforators", *J Xiangya Med*, 3:16 (2018).
77. N Pereira, D Valenzuela, G Mangelsdorff, et al. "Detection of perforators for free flap planning using smartphone thermal imaging: A concordance study with computed tomographic angiography in 120 perforators", *Plast Reconstr Surg*, 141:787–792 (2018).
78. GG Hallock, "Doppler sonography and color duplex imaging for planning a perforator flap", *Clin Plast Surg*, 30:347–357 (2003).
79. D Chubb, WM Rozen, IS Whitaker, MW Ashton, "Digital thermographic photography ("thermal imaging") for preoperative perforator mapping", *Ann Plast Surg*, 66: 324–325 (2011).
80. R Vardasca, EFJ Ring, P Plassmann, C Jones, "Thermal symmetry on extremities of normal subjects", *Thermol Int*, 17:19–23 (2007).
81. A Gatt, C Formos, K Cassar, et al. "Thermographic patterns of the upper and lower limbs: Baseline data", *Int J Vasc Med*, 2015:831369 (2015).
82. A Macdonald, N Petrova, S Ainarkar, et al. "Thermal symmetry of healthy feet: A precursor to a thermal study of diabetic feet prior to skin breakdown", *Physiol Meas*, 38(1):33–44 (2017).
83. DG Armstrong, LA Lavery, PJ Liswood, et al. "Infrared dermal thermography for the high-risk diabetic foot", *Phys Ther*, 77(2):169–175 (1997).
84. S Bagavathiappan, T Saravanan, J Philip, et al. "Infrared thermal imaging for detection of peripheral vascular disorders", *J Med Phys*, 34(1):43–47 (2009).
85. SJ Benbow, AW Chan, DR Bowsher, et al. "The prediction of diabetic neuropathic plantar foot ulceration by liquid-crystal contact thermography", *Diabetes Care*, 17(8):835–839 (1994).
86. M Midttun, P Sejrsen, WP Paaske, "Blood flow rate during orthostatic pressure changes in the pulp skin of the first toe", *Eur J Vasc Endovasc Surg*, 13(3):278–284 (1997).
87. M Midttun, P Sejrsen, WP Paaske, "Peripheral blood flow rates and microvascular responses to orthostatic changes in claudicants before and after revascularisation", *Eur J Vasc Endovasc Surg*, 17(3):225–229 (1999).
88. CJ Abularrage, AN Sidawy, G Aidinian, et al. "Evaluation of the microcirculation in vascular disease", *J Vasc Surg*, 42(3):574–581 (2005).
89. JM Cavaillon, "Once upon a time, inflammation", *J Venom Anim Toxins Incl Trop Dis*, 9(27):e20200147 (2021).
90. S Tejiram, SP Tranchina, TE Travis, JW Shupp, "The first 24 hours: Burn shock resuscitation and early complications", *Surg Clin North Am*, 103(3):403–413 (2023).
91. JL Ramirez-GarciaLuna, K Rangel-Berridi, R Bartlett, et al. "Use of infrared thermal imaging for assessing acute inflammatory changes: A case series", *Cureus*, 14(9):e28980 (2022).
92. M Schiavenato, RG Thiele, "Thermography detects subclinical inflammation in chronic tophaceous gout", *J Rheumatol*, 39(1):182–183 (2012).

93. JL Ramirez-GarciaLuna, SC Wang, T Yangzom, et al. "Use of thermal imaging and a dedicated wound-imaging smartphone app as an adjunct to staging hidradenitis suppurativa", *Br J Dermatol*, 186(4):723–726 (2022).
94. SJ Benbow, AW Chan, DR Bowsher, et al. "The prediction of diabetic neuropathic plantar foot ulceration by liquid-crystal contact thermography", *Diabetes Care*, 17(8): 835–839 (1994).
95. N Kaabouch, WC Hu, Y Chen, et al. "Predicting neuropathic ulceration: Analysis of static temperature distributions in thermal images", *J of Biomed Opt*, 15(6): 061715 (2010).
96. LA Lavery, KR Higgins, DR Lanctot, et al. "Home monitoring of foot skin temperatures to prevent ulceration", *Diabetes Care*, 27:2642–2647 (2004).
97. RG Frykberg, IL Gordon, AM Reyzelman, et al. "Feasibility and efficacy of a smart mat technology to predict development of diabetic plantar ulcers", *Diabetes Care*, 40:973–980 (2017).
98. T Nagase, H Sanada, K Takehara, et al. "Variations of plantar thermographic patterns in normal controls and non-ulcer diabetic patients: Novel classification using angiosome concept", *J of Plastic, Reconstructive & Aesthetic Surgery: JPRAS*, 64(7): 860–866 (2011).
99. M Bharara, J Schoess, A Nouvong, DG Armstrong, "Wound inflammatory index: A 'proof of concept' study to assess wound healing trajectory", *J Diabetes Sci Technol*, 4(4): 773–779 (2010).
100. LA Lavery, BJ Petersen, DR Linders, et al. "Unilateral remote temperature monitoring to predict future ulceration for the diabetic foot in remission", *BMJ Open Diab Res Care*, 7:e000696 (2019).
101. TC Carpenter, S Schomberg, KR Stenmark, "Endothelin-mediated increases in lung VEGF content promote vascular leak in young rats exposed to viral infection and hypoxia", *Am J Physiol Lung Cell Mol Physiol*, 289(6):L1075–L1082 (2005).
102. TE Serena, JR Hanft, R Snyder, "The lack of reliability of clinical examination in the diagnosis of wound infection: Preliminary communication", *Int J Low Extrem Wounds*, 7(1):32–35 (2008).
103. T Swanson, K Ousey, E Haesler, et al. "IWII wound infection in clinical practice consensus document: 2022 update", *J Wound Care*, 31(Sup12):S10–S21 (2022).
104. KY Woo, RG Sibbald, "A cross-sectional validation study of using NERDS and STONEES to assess bacterial burden", *Ostomy Wound Manage*, 55(8):40–48 (2009).
105. O Rahbek, HC Husum, M Fridberg, et al. "Intrarater reliability of digital thermography in detecting pin site infection: A proof of concept study", *Strategies Trauma Limb Reconstr*, 16(1):1–7 (2021).
106. A Chanmugam, D Langemo, K Thomason, et al. "Relative temperature maximum in wound infection and inflammation as compared with a control subject using long-wave infrared thermography", *Adv Skin Wound Care*, 30(9):406–414 (2017).

107. M Oe, RR Yotsu, H Sanada, et al. "Thermographic findings in a case of type 2 diabetes with foot ulcer and osteomyelitis", *J Wound Care*, 21(6): 274–278 (2012).
108. G Saiko, "Skin temperature: The impact of perfusion, epidermis thickness, and skin wetness", *Appl Sci*, 12: 7106 (2022).
109. RP Cole, PG Shakespeare, HG Chissell, SG Jones, "Thermographic assessment of burns using a nonpermeable membrane as wound covering", *Burns*, 17(2): 117–122 (1991).
110. DG Moreira, JT Costello, CJ Brito, et al. "Thermographic imaging in sports and exercise medicine: A Delphi study and consensus statement on the measurement of human skin temperature", *J Therm Biol*, 69: 155–162(2017).
111. JH Samies, M Gehling, TE Serena, RA Yaakov, "Use of a fluorescence angiography system in assessment of lower extremity ulcers in patients with peripheral arterial disease: A review and a look forward", *Seminars in Vascular Surgery*, 28(3–4): 190–194 (2015).
112. K Igari, T Kudo, H Uchiyama, et al. "Indocyanine green angiography for the diagnosis of peripheral arterial disease with isolated infrapopliteal lesions", *Annals of Vascular Surgery*, 28(6): 1479–1484(2014).
113. A Refaat, ML Yap, G Pietersz, et al. "In vivo fluorescence imaging: Success in preclinical imaging paves the way for clinical applications", *J Nanobiotechnol* 20: 450 (2022).
114. NA Wisniewski, SP Nichols, SJ Gamsey, et al. "Tissue-integrating oxygen sensors: Continuous tracking of tissue hypoxia", *Adv Exp Med Biol*, 977: 377–383 (2017).
115. SP Nichols, MK Balaconis, RM Gant, et al. "Long-term in vivo oxygen sensors for peripheral artery disease monitoring", *Adv Exp Med Biol*, 1072: 351–356(2018).
116. B Kjeldstad, T Christensen, A Johnsson, "Porphyrin photosensitization of bacteria", *Adv Exp Med Biol*, 193: 155–159 (1985).
117. YS Cody, DC Gross, "Characterization of pyoverdin (pss), the fluorescent siderophore produced by Pseudomonas syringae pv. Syringae", *Appl Environ Microbiol*, 53: 928–934 (1987).
118. L Le, M Baer, P Briggs, et al. "Diagnostic accuracy of point-of-care fluorescence imaging for the detection of bacterial burden in wounds: Results from the 350-Patient FLAAG trial", *Adv Wound Care*, 10(3):123–136 (2021).
119. R Hill, KY Woo, "A prospective multi-site observational study incorporating bacterial fluorescence information into the UPPER/ LOWER wound infection checklists", *Wounds*, 32: *299–308 (2020)*.
120. TE Serena, K Harrell, L Serena, RA Yaakov, "Real-time bacterial fluorescence imaging accurately identifies wounds with moderate-to-heavy bacterial burden", *J Wound Care*, 28: 346–357 (2019).
121. K Ottolino-Perry, E Chamma, KM Blackmore, et al. "Improved detection of clinically relevant wound bacteria using autofluorescence image-guided sampling in diabetic foot ulcers", *Int Wound J*, 14: 833–841 (2017).

122. LM Jones, D Dunham, MY Rennie, et al. "In vitro detection of porphyrin-producing wound bacteria with real-time fluorescence imaging", *Future Microbiol*, 15: 319–332 (2020).
123. R Raizman, D Dunham, L Lindvere-Teene, et al. "Use of a bacterial fluorescence imaging device: Wound measurement, bacterial detection and targeted debridement", *J Wound Care*, 28: 824–834 (2019).
124. R Hill, MY Rennie, J Douglas, "Using bacterial fluorescence imaging and antimicrobial stewardship to guide wound management practices: A case series", *Ostomy Wound Manag*, 64: 18–28 (2018).
125. R Raizman, "Fluorescence imaging guided dressing change frequency during negative pressure wound therapy: A case series", *J Wound Care*, 28: S28–S37 (2019).
126. B. Aung, "Can fluorescence imaging predict the success of CTPs for wound closure and save costs?" *Today's Wound Clin*, 13: 22–25 (2019).
127. The Delphi method is an anonymous process used to collect knowledge from a panel of experts and then subsequently distill opinions to a point of high consensus. Experts respond to iterative rounds of questionnaires, and the responses are aggregated based on concurrence of opinion.
128. AR Oropallo, C Andersen, R Abdo, et al. "Guidelines for point-of-care fluorescence imaging for detection of wound bacterial burden based on Delphi consensus", *Diagnostics (Basel)*, 11(7):1219 (2021).
129. L Le, M Baer, P Briggs, et al. "Diagnostic accuracy of point-of-care fluorescence imaging for the detection of bacterial burden in wounds: Results from the 350-patient fluorescence imaging assessment and guidance trial", *Adv Wound Care*, 10:123–136 (2021).
130. S Rahm, J Woods, S Brown, et al. "The use of point-of-care bacterial autofluorescence imaging in the management of diabetic foot ulcers: A pilot randomized controlled trial", *Diabetes Care*, 45(7):1601–1609 (2022).
131. LA Lavery, DG Armstrong, RP Wunderlich, et al. "Risk factors for foot infections in individuals with diabetes", *Diabetes Care*, 29:1288–1293 (2006).
132. MC Robson, "Wound infection a failure of wound healing caused by an imbalance of bacteria", *Surgical Clinics of North America*, 77(3):637–650 (1997).
133. R Edwards, K Harding, "Bacteria and wound healing", *Current Opinion in Infectious Diseases*, 17(2):91–96 (2004).
134. AC Brone, M Vearncombe, RG Sibbald, "High bacterial load in asymptomatic diabetic patients with neurotrophic ulcers retards wound healing after application of dermagraft",*Ostomy Wound Management*, 47(10):44–49 (2001).
135. MF Yoder, WS Kisaalita, "Fluorescence of pyoverdin in response to iron and other common well water metals", *J Env ScieHealth, Part A*, 41(3): 369–380 (2006).
136. W Groner, JW Winkelman, AG Harris, et al. "Orthogonal polarization spectral imaging: A new method for study of the microcirculation", *Nat Med*, 5(10): 1209–1212 (1999).

137. O Goertz, A Ring, A Kohlinger, et al. "Orthogonal polarization spectral imaging: A tool for assessing burn depth?" *Ann Plast Surg*, 64(2): 217–221 (2010).
138. CE Virgini-Magalhaes, CL Porto, FF Fernandes, et al. "Use of microcirculatory parameters to evaluate chronic venous insufficiency", *J Vasc Surg*, 43(5): 1037–1044 (2006).
139. PT Goedhart, M Khalilzada, R Bezemer, et al. "Sidestream Dark Field (SDF) imaging: A novel stroboscopic LED ring-based imaging modality for clinical assessment of the microcirculation", *Opt Exp*, 15(23): 15101 (2007).
140. H Sherman, S Klausner, WA Cook, "Incident dark-field illumination: A new method for microcirculatory study", *Angiology*, 22:295–303 (1971).
141. E Gilbert-Kawai, J Coppel, V Bountziouka, et al. "A comparison of the quality of image acquisition between the incident dark field and sidestream dark field video-microscopes", *BMC Med Imaging* 16: 10 (2016).
142. I Schelkanova, A Pandya, G Saiko, et al. "Spatially resolved, diffuse reflectance imaging for subsurface pattern visualization toward development of a lensless imaging platform: Phantom experiments", *J Biomed Opt*, 21(1): 015004 (2016).
143. A Pandya, I Schelkanova, A Douplik, "Spatio-angular filter (SAF) imaging device for deep interrogation of scattering media", *Biomed Opt Expr*, 10(9): 4656–4663 (2019).
144. S Lange-Asschenfeldt, A Bob, D Terhorst, et al. "Applicability of confocal laser scanning microscopy for evaluation and monitoring of cutaneous wound healing", *J Biomed Opt*, 17(7): 076016 (2012).
145. M Rajadhyaksha, S Gonzalez, JM Zavislan, et al. "In vivo confocal scanning laser microscopy of human skin II: Advances in instrumentation and comparison with histology", *J Invest Derm*, 113(3): 293–303 (1999).
146. AA Altintas, M Guggenheim, MA Altintas, et al. "In vivo confocal laser scanning microscopy imaging of skin inflammation: Clinical applications and research directions", *J Burn Care & Res*, 30(6): 1007–1012 (2009).
147. M Ghijsen, TB Rice, B Yang, et al. "Wearable speckle plethysmography (SPG) for characterizing microvascular flow and resistance," *Biomed Opt Express*, 9: 3937–3952 (2018).
148. MK Razavi, DPT Flanigan, SM White, TB Rice, "A real-time blood flow measurement device for patients with peripheral artery disease", *J Vasc Interven Rad*, 32(3):453–458 (2021).
149. AA Kamshilin, VV, Zaytsev, AV Lodygin, et al. "Imaging photoplethysmography as an easy-to-use tool for monitoring changes in tissue blood perfusion during abdominal surgery", *Sci Rep*,12: 1143 (2022).
150. T Burton, G Saiko, A Douplik, "Feasibility study of remote contactless perfusion imaging with consumer-grade mobile camera", *Adv Exp Med Biol*, 1395:289–293 (2022).
151. SP Schraven, B Kossack, D Strüder, et al. "Continuous intraoperative perfusion monitoring of free microvascular anastomosed fasciocutaneous flaps using remote photoplethysmography", *Sci Rep*, 13: 1532 (2023).

152. MF la Cour, S Mehrvar, JS Heisner, et al. "Optical metabolic imaging of irradiated rat heart exposed to ischemia-reperfusion injury", *J Biomed Opt*, 23(1): 1–9 (2018).
153. B Chance, B Schoener, R Oshino, et al. "Oxidation-reduction ratio studies of mitochondria in freeze-trapped samples. NADH and flavoprotein fluorescence signals", *J Biol Chem*, 254(11): 4764–4771 (1979).
154. BR Masters, A Kriete, J Kukulies,"Ultraviolet confocal fluorescence microscopy of the in vitro cornea: redox metabolic imaging", *Appl Opt*, 32: 592–596 (1993).
155. SH Huang, AA Heikal, WW Webb, "Two-photon fluorescence spectroscopy and microscopy of NAD(P)H and flavoprotein", *Biophys J*, 82: 2811–2825 (2002).
156. TS Blacker, ZF Mann, JE Gale, et al. "Separating NADH and NADPH fluorescence in live cells and tissues using FLIM", *Nat Commun*, 5: 3936 (2014).
157. S Mehrvar, KT Rymut, FH Foomani, et al."Fluorescence imaging of mitochondrial redox state to assess diabetic wounds", *IEEE J Transl Eng Health Med*, 18(7):1800809 (2019).
158. OI Kolenc, KP Quinn, "Evaluating cell metabolism through autofluorescence imaging of NAD(P)H and FAD", *Antioxid Redox Signal*, 30(6):875–889 (2019).

CHAPTER 6

Therapeutic Approaches

ABOUT HALF OF ALL foot ulcers are clinically infected at the time the patient presents to a clinician, and ulcers are the most frequent predisposing factor for foot infections. These foot infections may begin superficially, but if left untreated, they can spread to the contiguous subcutaneous tissues. Ultimately, the infectious process may involve muscles, tendons, bones, and joints. These deep infections are potentially disastrous and can rapidly progress to septic gangrene, which may eventually require amputation. That is why timely treatment is of paramount importance in DFU management.

DFU treatment is different at various stages of the disease. Armstrong and Lipsky[1] proposed a stepwise approach to treating diabetic foot infections:

> Most patients must first be medically stabilized, and any metabolic problems should be addressed. Antibiotic therapy is not required for uninfected wounds but should be carefully selected for all infected lesions. Initial therapy is empiric but may be modified according to the culture and sensitivity results and the patient's clinical response. Surgical intervention is usually required in cases of retained purulence or advancing infection despite optimal medical therapy

In addition to wound dressings and antibiotic regimens, there are numerous additional methods for DFU treatment such as surgical methods and wound therapies, which include Negative Pressure Wound Therapy

(NPWT), hyperbaric oxygen, ultrasound, and Low-Level Laser Therapy (LLLT). In this multitude of therapeutic approaches, some methods are light-based. A brief overview of these light-based methods will be the subject of this chapter.

Light can produce multiple effects on the tissue. These effects can be classified based on light intensity into thermal and nonthermal ones. As these effects differ, we will consider them separately and refer to nonthermal (low intensity) light as limited by intensities of <110 mW/cm². Unlike high-power light, photochemical reactions are the primary means of interacting low-intensity light with tissue. Using existing terminology, we will refer to these therapies as phototherapy or Low-Level Laser Therapy, or LLLT.

6.1 SURGICAL METHODS

Invasive surgery is often the last resort for diabetic foot ulcers unresponsive to less aggressive care. Surgical intervention is usually required in cases of retained purulence or advancing infection despite optimal medical therapy. These measures include proper surgical debridement, drainage, and wound lavage. For patients whose wounds do not heal after adequate treatment, the best long-term outcome may be achieved through revascularization or, in some cases, amputation.[1]

According to the CDC, at least 60% of non-traumatic lower limb amputations occur among people with diabetes.[2] In the US in 2018, there were 154,000 lower-extremity amputations (6.1 per 1,000 adults with diabetes).[3] Until recently, amputation was the only option in serious cases. However, recently, revascularization gained traction as a limb-salvage alternative.

6.1.1 Wound Cleaning

Necrotic tissue, excessive bacterial burden, and cellular debris can inhibit wound healing. According to Whitney et al.[4] 'initial debridement is required to remove the obvious necrotic tissue, excessive bacterial burden, and cellular burden of dead and senescent cells. Maintenance debridement is needed to maintain the appearance and readiness of the wound bed for healing.' Surgical debridement can be performed by a sharp instrument (scissors, scalpels, forceps) or lasers.

Debridement also can be performed using lasers. Laser debridement is based on the controlled vaporization of the superficial layers of the wound bed. Unlike other methods dependent on the clinician's manual control, laser debridement is electronically controlled, improving precision and reducing the risk of healthy tissue damage. In particular, the laser type and the number of passes performed determine the depth of tissue ablation.[5]

Wound debridement using lasers began in the 1970s, with the successful report of a continuous-beam Carbon Dioxide (CO_2) laser used for skin graft preparation of infected decubitus ulcers.[6] Since then, laser debridement has been demonstrated for burns[7,8] and wound management.

In a study on 164 patients (82 in the CO_2 laser group and 82 matched in the routine group), Guan et al.[9] found that the time-to-heal for patients in the CO_2 laser group (41.30 ± 17.11) was significantly shorter than that of the patients in the routine group (48.51 ± 24.32) (p = 0.015).

Initial works were performed using surgical CO_2 lasers. Recently, the focus changed to Erbium:Yttrium-Aluminum-Garnet (Er:YAG) lasers. Since Er:YAG laser energy has greater than 12 times more water absorption efficiency than CO_2 lasers, water in the tissue is rapidly expanded to eject the charred debris from the wound surface without leaving behind a necrotic eschar. For example, Er:YAG laser debridement was studied in conjunction with sharp debridement in patients with venous ulcers and DFU.[10] Post-laser debridement prevalence of bacterial load >10^5 CFU/g decreased from 72.7% to 40.9%.

6.1.2 Wound Closure

Some thermal and photochemical methods in wound healing aim to improve traditional closure methods (sutures, staples, and clips). These methods include laser tissue welding (using a laser thermal effect to bond tissue together) and laser tissue welding with protein solders.[11] It is deemed that photochemical tissue welding[12] provides the strongest bonds between all these methods. For example, the Kochevar group developed and successfully applied a light-activated method (Photochemical Tissue Bonding or PTB) for closing surgical incisions in the cornea[13] and wounds[14] using green light (532nm) and a photosensitizing dye (Rose Bengal – RB).

6.2 LASER ANTISEPTICS

An antiseptic is a substance or method that prevents the growth of microorganisms. In general, any chronic wound is considered contaminated with bacteria. A wound containing contaminated foci with more than 10^5 organisms per gram of tissue cannot be readily closed, as the incidence of wound infection that follows is 50–100%.[15] However, in some cases (diabetes mellitus, anemia, poor immunity, etc.), the infection can develop in a wound with an even smaller bacterial load.

Numerous antiseptic methods exist: mechanical (surgical), physical, chemical, and biological (e.g., antibiotics, bacteriophages). Among them, there are several light-based approaches to managing bacterial load. They include thermal (debridement, where a surgical laser evaporates necrotic tissues) and photochemical methods.

Debridement was discussed in Section 6.1. This section will focus on photochemical methods, which can be split into UV-C, Photodynamic Therapy, and phototherapy. The term phototherapy in the context of this section refers to light therapy without using an external agent (like a photosensitizer in PDT).

6.2.1 UV-C

It is well known that microorganisms are inactivated with ultraviolet light (UV-C – at wavelengths of 240–280 nm, the so-called germicidal range) due to their ability to cause DNA damage and induce sublethal damage. For example, UV-C demonstrates inhibitory action on antibiotic-resistant strains of bacteria.[16] In particular, Coohill and Sagripanti considered the mechanisms of action of 254 nm UV-C.[17] However, light in this range can also damage the host tissue. While UV-C is widely used for disinfection tools and unoccupied spaces, using it in vivo is quite problematic.

Recently, a new UV range (207–222 nm) was proposed to inactivate airborne infections like COVID-19.[18] It is suggested that these wavelengths do not cause the human health issues associated with direct exposure to conventional germicidal UV light, as their penetration depth in biological tissues is less than a few micrometers. Thus, it cannot reach living human cells in the skin or eyes, being absorbed in the skin stratum corneum or the ocular tear layer. But, it is sufficient to kill microorganisms in the air or on the surface. However, whether it can be used as an in vivo antiseptic is unclear.

6.2.2 Photodynamic Therapy

Photodynamic Therapy (PDT) refers to the therapeutic use of photochemical reactions mediated through the interaction of photosensitizing agents, light, and oxygen. PDT is a two-step process. In step one, the light-sensitive agent (named Photosensitizer, or PS) is administered to the patient by one of several routes (e.g., topical, oral, intravenous) and allowed to be taken up by the target cells. The second step involves the activation of the photosensitizer with a specific wavelength of light directed toward the target tissue. The application of PDT is being explored in many fields, including oncology, dermatology, cardiovascular, and ophthalmology.

The term 'photodynamic' was coined in 1907 by Herman von Tappeiner, who, in 1904, together with dermatologist A. Jesionek performed the first PDT on a patient with skin carcinoma using topically applied eosin as PS along with white light.

PDT's antibacterial properties were discovered in 1900 by the same group when Tappeiner's student Oscar Raab noticed that dyes like acridine, along with light, can kill paramecia.[19] However, while antimicrobial properties of PDT were discovered even earlier than anti-tumor ones, the focus for PDT use in the next 100 years was primarily in oncology.

The interest in the antimicrobial use of PDT as an alternative to antibiotics was resurrected in the 2000s with the growth of Antimicrobial Resistance (AMR).

The antimicrobial use of PDT is oftentimes referred to as antibacterial Photodynamic Therapy (aPDT). aPDT was introduced in 1966 by Macmillan et al.[20] who used Toluidine blue against microorganisms like bacteria, algae, and yeast. It was observed that 99% of bacteria were killed within 30 min of exposure to 21–30 mW of light at 632.8 nm from a continuous-wave gas laser.

There is growing evidence that aPDT is efficient in eliminating bacteria, fungi, and viruses.[21] In the first demonstration of the aPDT approach to destroy bacteria in the wound, Hamblin et al.[22] used targeted polycationic photosensitizer combined with a 660-nm laser to kill *Escherichia coli* infection in mice rapidly. In a study by Morley et al.[23] PDT was explored as a novel means of treating chronic ulcers (specifically leg ulcers) and overcoming the antibiotic-resistant nature many ulcers exhibit. Comparing the effects to the placebo condition, patients treated with a light-sensitive PPA904 drug and light had a reduction in bacterial load post-dose ($p<0.001$). Following three months, half of the patients in the sample set had their ulcers completely healed, compared with eight patients on the placebo conditions. This study, however, utilized a very small sample size of eight patients; further investigation is needed to confirm and substantiate the results.

6.2.2.1 Mechanism of Action

A ground state of PS absorbs photons and is transformed into its excited state (PS*), followed by the photophysical pathways of internal conversion and intersystem crossing to the lowest excited singlet state (^{1}PS*) or lowest excited triplet state (^{3}PS*). These lowest excited states of PS* can directly revert to the ground state by radiation emission. The lowest excited triplet state ^{3}PS* has a longer lifetime for energy transfer to biological substrates to form Reactive Oxygen Species (ROS) (Type I reaction) or electron transfer to oxygen to form the singlet oxygen ($^1O_2$) (Type II reaction).[24] Type I reaction involves electron transfer from triplet state PS to an organic substrate within the cells, producing free radicals. In a Type II reaction, energy transfer occurs between the excited PS and the ground-state molecular oxygen, producing singlet oxygen that can interact with a large number of molecules in the cell to generate oxidized products. Type II reaction is the primary mechanism of PDT in cancer therapy. Generally, the proportion of Type I/Type II reactions depends on the PS used for PDT.

6.2.2.2 Photosensitizers

Antimicrobial photosensitizers are very different from anti-tumor ones. For example, it was established that cationic dyes are more effective against bacteria than anionic[25] or neutral PS. In particular, cationic molecules carry a positive charge on their functional groups, so they are easily bound to and taken up by bacteria that possess a negative charge on the surface.

In addition, antimicrobial PS is being applied topically. Thus, a broad spectrum range from UV to infrared can be potentially used for photoexcitation. It differs from anti-tumor PS, which needs to be photoactivated deep in the tissue, limiting the light wavelength options to red and infrared light with a deeper penetration depth (see Appendix B for further details).

The antimicrobial photosensitizers were reviewed by Ghorbani et al.[26] They categorized them into four groups: synthetic dyes, natural PSs, tetrapyrrole structures, and nanostructures.

6.2.2.2.1 Synthetic Dyes

Phenothiazinium dyes are the most prominent sub-group of synthetic dyes for aPDT. Among them, the most commonly used dyes are Methylene blue (Mb) and Toluidine blue (Tb), which due to their cationic nature, bind to most gram-positive and gram-negative microorganisms with high affinity and, therefore, are primary PS for aPDT in clinical settings.

6.2.2.2.2 Tetrapyrroles

Tetrapyrroles are one of the largest and first introduced PSs groups. In particular, they include porphyrins, the most well-known photosensitizers.

Some bacteria, like anaerobic bacteria *Propionibacterium acnes* and *Bacteroides species*, and oral bacteria, including *Porphyromonas gingivalis*,

Prevotella spp, and *Aggregatibacter actinomycetemcomtans*, which produce black pigment, produce large amounts of porphyrin.[27,28,29]

The tetrapyrroles group also includes Zn phthalocyanine derivatives[30] and chlorine.[31]

6.2.2.2.3 Natural PS

Natural photosensitizers refer to compounds extracted from plants and other organisms. They include coumarins, furanocoumarins, benzofurans, anthraquinones, and flavin derivatives.

Hypericum perforatum, or St John's-wort, is a flowering plant traditionally known for its healing effects on burns and skin injuries. Hypericin is an anthraquinone derivative isolated from *Hypericum perforatum*. Garcia et al.[32] found that hypericin is an effective aPDT PS against gram-positive bacteria, but there is no effect against gram-negative ones.

Curcumin is a natural PS isolated from the root of a plant called *Curcuma longa*, known for its anti-oxidant, antibacterial, anti-inflammatory, and wound-healing properties. Curcumin was found to be 300 times more effective against the gram-positive *S. aureus* than the gram-negative *E.coli* and *Salmonella typhimurium*.[33]

The most promising natural PS are flavin derivatives,[34] demonstrating a cationic nature and high affinity to gram-positive and gram-negative microorganisms.

6.2.2.2.4 Nanostructures

Research on using nanoparticles as PS is focused primarily on anti-cancer treatments. However, some antimicrobial applications have also been explored.

According to Perni et al.[35] nanostructures as PS can be classified into four groups: PS embedded in nanoparticles, PS bound to the surface of nanoparticles, PS-accompanying nanostructures, and nanoparticles as PS.

Most nanoparticles are used as delivery vehicles for PSs, such as tetrapyrroles, natural products, and phenothiazinium dyes. For example, Tsai et al.[36] tested the antimicrobial activity of hematoporphyrin (Hp) enclosed in either liposomes or micelles on gram-positive pathogens such as MSSA, MRSA, *S. epidermidis*, and *Streptococcus pyogenes*. The results indicated that aPDT with Hp encapsulated in micelles was more effective than the one encapsulated in liposomes at the same Hp doses.

PS can also be bound to the surface of nanoparticles to enhance the antimicrobial properties of PS. For example, porphyrin has a tendency for carbon nanotubes,[37] while Toluidine blue tends to bind to the surface of gold nanoparticles.[38]

Fullerenes and semiconductor materials (like TiO_2 and ZnO) can act as PS. Studies on *E. coli* in vitro showed that cationic fullerene N,N-dimethyl-2-(40-N,N,N-trimethyl-aminophenyl) fulleropyrrodinium iodide ($DTC60_{2+}$) hindered *E. coli* proliferation about 3.5 $\log_{10}$ after 30 min of under white light exposure compared to the negligible killing effect of non-charged N-methyl-2-(40-acetamidophenyl) fulleropyrrolidine (MAC 60).[39]

6.2.2.2.5 Prodrugs

An alternative approach to direct applications of photosensitizers is to use prodrugs, which are converted into photosensitizers (typically, porphyrins). It can be considered a 'Trojan horse' approach. For example, 5-Aminolevulinic Acid (ALA) and Methyl Aminolevulinate (MAL) can be converted into the Photoactive protoporphyrin IX (PpIX) via the heme biosynthetic pathway. This approach is a primary PDT tactic in dermatology and is used to treat various skin conditions.[40] For example, the US FDA approved 5-ALA 20% solution and 10% nanoemulsion for treating nonhyperkeratotic Actinic Keratosis (AK) of the face and scalp.

This approach can be used for antimicrobial purposes as well. For example, it is well established that 5-ALA is metabolized by multiple bacteria strains into porphyrins, which can be used to target metabolically active colonies. In particular, Lee et al.[41] demonstrated that *Proteus hauseri, Aeromonas hydrophila, Bacillus cereus,* and *Staphylococcus aureus* were killed via antimicrobial PDT with a 1% 5-ALA and reached 100% killing rate at optimal conditions.

6.2.3 Phototherapy

It has been documented that visible light, especially in the wavelength 400–500 nm (blue light)[42] and NIR[43] at specific energy doses, affects the growth of some fungal and bacterial species. However, the technology has not been translated into medical practice yet as it has only been explored using in vitro or in animal models.

6.2.3.1 *Blue Light*

The antimicrobial properties of blue light phototherapy were reviewed by Percival et al.[44] It has been found that the research is predominantly being conducted in vitro or in animal models. Papageorgiou et al.[45] divided 107 patients with acne vulgaris into four treatment groups: blue light, white light, mixed blue, and red light, and 5% benzoyl peroxide cream. The mixed blue and red light phototherapy efficacy was significantly superior to the other three treatments. After 12 weeks, mixed blue and red light phototherapy achieved a 76% improvement in inflammatory lesions.

It should be noted that antibacterial phototherapy is associated with a much higher light dose than LLLT therapies. In general, low energy doses (<30 J/cm^2) seem to enhance bacterial proliferation, while higher doses must be applied to achieve a germicidal effect. Particularly, for the gram-negative *P. aeruginosa* and *E. coli*, a dose of 180 J/cm^2 at 405 nm is needed.[46] As such, there are certain concerns about the usefulness of blue light phototherapy. In particular, in,[47] the authors found that blue light reduces keratinocyte proliferation and migration at high doses and, therefore, could negatively affect wound healing.

While the exact mechanism of bacterial inactivation by blue light phototherapy has not been established, the predominant hypothesis is that the damage to microbial cells is associated with generating Reactive Oxygen Species (ROS).

ROS can be generated in photochemical reactions in several ways. For example, Hockberger et al.[48] showed that blue light stimulated H_2O_2 production in cultured mouse (3T3 fibroblasts), monkey (kidney epithelial cells), and human (foreskin keratinocytes) cells.

The related mechanism can be associated with photoexciting porphyrins that can act as endogenous photosensitizers within the bacteria.[49,50] It is known that multiple bacteria produce high levels of coproporphyrins, which can become photosensitized by blue light and produce Reactive Oxygen Species, causing cell death.[51] As a result, PDT for the treatment of Acne Vulgaris caused by *Propionibacterium acnes* can be performed even without photosensitizer administration. However, off-label use of prodrugs (5-ALA and MAL, see below) to increase porphyrin production is the standard practice in clinical trials.[52]

6.2.3.2 NIR

The mechanisms of bacterial inactivation by NIR phototherapy are not understood. For example, it can be attributed to a stimulatory effect on the innate and adaptive immune system.[53]

In a recent review, Percival et al.[54] concluded that the exposure to red and infrared light (700–1000 nm) at doses ranging from 1 to 120 J/cm^2

seems not to have a significant effect on the viability of gram-positive (*S. aureus*) and gram-negative (*P. aeruginosa* and *E. coli*) bacteria in planktonic or biofilm states.

6.3 LASER SYSTEMIC THERAPY

The use of light in medicine has a long history. Ancient Egyptians and Greeks believed that the sun could strengthen and heal the body.[55] In the Middle Ages, sunlight was also considered to be an ally in the battle against virulent diseases such as the plague.[56]

The modern applications of light therapies started with the works of Niels Ryberg Finsen, who is considered to be the father of phototherapy. He devised the treatment of small-pox in red light (1893) and lupus (1895), which in 1903 was awarded a Nobel Prize in Medicine:'in recognition of his work on the treatment of diseases, and in particular the treatment of lupus vulgaris by means of concentrated light rays.' Finsen also suggested using general sunbathing to treat tuberculosis, which resulted in the successful (and popular) treatment of surgical tuberculosis at high elevations in Switzerland.

Light therapy applications were broadened significantly with the invention of lasers in the 1960s. Although light therapy is mostly applied to localized diseases, its effect is often considered to be restricted to the exposed area. There are reports of systemic effects of light therapy acting at a site distant from the illumination.[57,58] It is well known that UV light can have systemic effects,[59] and it is expected that red and NIR light can have systemic effects, likely to be mediated by soluble mediators such as endorphins and serotonin. In addition, systemic effects can be achieved through laser acupuncture[60] (the stimulation of specific acupuncture points by a focused laser beam) or through laser radiation of blood.[61]

6.4 LOW-LEVEL LOCAL BIOSTIMULATION

Since the invention of lasers, it has been known that low levels of visible or Near-Infrared (NIR) light can reduce pain and inflammation, promote the healing of wounds, deeper tissues, and nerves, as well as prevent tissue damage through a stimulatory effect (biostimulation).[62].

Early sources in wound healing were HeNe and argon lasers. Today, while HeNe lasers are still used, most work is done with GaAs and GaAlAs diodes, with wavelengths between 820 nm and 904 nm. Early investigators established 1–4 J/cm^2 dosages, and these remain the most frequently used radiant exposures delivered to treatment sites.

In addition to lasers (noble gas and semiconductor), it was found that Light-Emitting Diodes (LED), Superluminuos Diodes (SLD), and polarized light also demonstrate therapeutic effects. Thus, an array of different (and often confusing) terms evolved: phototherapy, biostimulation, photobiomodulation, Low-Level Laser Therapy (LLLT), Low-Power Laser Therapy (LPLT), Low-Intensity Laser Therapy (LILT), cold laser, soft laser, therapeutic laser, Light-Emitting Diode, low reactive level laser, and diode laser. Therefore, the terms' phototherapy,' or 'light therapy' have been recommended.[63]

Phototherapy demonstrates several effects that promote wound healing, including wound closure and antimicrobial properties. While these effects are intervened (namely, bacterial load prevents wound closure), we will focus here on wound closure as we discuss the antimicrobial properties of low-intensity light in Section 6.2.

6.4.1 Clinical Data

Clinical data on using LLLT for wound healing is scarce. As we found just 4 DFU-specific trials, we will also include information for trials for wounds with other etiologies (namely, pressure injury).

In particular, Schindl et al.'s[64] double-blinded placebo-controlled trial reports that treatment using HeNe laser increased skin temperature in patients with diabetic microangiopathy. Design faults include a small sample size and a lack of long-term follow-up. The dose of 30 J/cm^2 is higher than that used in other investigations, and the exposure time is unusually long at 50-min exposure, which can be impractical for clinical applications.

A 100% cure rate in 1.5–2 months, with a total resolution of seven chronic diabetic ulcers, is reported by Kazemi-Khoo.[65] The dosage used is consistent with that used by several other investigators, i.e., 4–6 J/cm^2. This dosage was also recommended for wound healing. However, the use of various application methods (i.e., topical, acupuncture, and intravenous) makes treatment effects difficult to compare. Despite the positive result, no evidence or conclusions can be drawn due to the small sample size and lack of a control group.

A randomized clinical trial to evaluate the efficacy of LLLT on wound healing was performed by Kajagar et al.[66] on 68 patients with chronic diabetic foot ulcers having negative cultures. Thirty-four patients were treated just with the conventional therapy (systemic antibiotics), and the remaining 34 patients with LLLT (2 or 4 J/cm^2) combined with the conventional therapy. By recording healing and percentage reduction in ulcer areas over 15 days, the authors demonstrated the beneficial effects of LLLT in treating diabetic foot ulcers.

Nussbaum et al.[67] performed a double-blind randomized trial with stratification for ulcer location to the buttock or lower extremity to compare the effects of UV-C with placebo-UV-C on pressure ulcer healing (stage 2–4 pressure ulcers) in individuals with spinal cord injury. Subjects were followed up for one year post-intervention. Results showed that UV-C was beneficial only for stage 2 buttock ulcers.

Lucas et al.[68] performed a prospective, observer-blinded multicenter randomized clinical trial to assess the efficacy of LLLT in treating stage 3 decubitus ulcers. A total of 86 patients were enrolled in the study. Treatment was the prevailing consensus decubitus treatment ($n = 47$); one group ($n = 39$) had LLLT, in addition, five times a week over a period of six weeks. During the treatment period, 11% of the patients in the control group and 8% of the patients in the LLLT group developed a stage 4 decubitus ulcer. The patients' Norton scores at six weeks did not change during the treatment period. This trial found no evidence of Low-Level Laser Therapy as an adjuvant to the consensus decubitus ulcer treatment.

Schubert[69] showed the efficacy of phototherapy on pressure ulcer healing in elderly patients (> or = 65 years) after a falling trauma. Phototherapy consisted of nine-minute treatments with pulsed monochromatic infrared (956 nm) and red (637 nm) light. The ulcer surface area was traced weekly. Patients treated with pulsed monochromatic light had a 49% higher ulcer healing rate and a shorter time to 50% and 90% ulcer closure than controls. Their mean ulcer area was reduced to 10% after five weeks compared with nine weeks for the controls.

Clinical data on phototherapy in wound healing were reviewed by Whinfield.[70] The authors analyzed 17 clinical studies and found that a positive treatment effect was recorded in 13 reports, while the remaining 4 found phototherapy treatment ineffective. The authors concluded that the results of clinical trials on wound healing by phototherapy have been inconclusive. The only consistent finding between numerous clinical trials and animal models is that the positive effect of phototherapy on wound healing is more profound for 'healing impaired' wounds such as diabetic ulceration.

6.4.2 Combined Methods

Phototherapy can be combined with other topical and adjunctive therapies, such as ultrasound, hyperbaric oxygen, and wound dressings. In particular, Nussbaum et al.[71] compared laser vs. ultrasound plus UV-C therapy and found a mean weekly reduction in wound surface area of 23.7% for laser and standard care vs.53.5% for ultrasound and UV-C plus standard care vs. 32.4% for standard care. In an uncontrolled study, Landau et al.[72] treated 374 subjects with chronic ulcers and reported successful treatment outcomes in 78% of the subjects. However, as hyperbaric oxygen was used in conjunction with a laser, it is unknown to which intervention the results can be attributed.

6.4.3 Mechanisms Involved in DFU Healing

A significant body of works (primarily based on in vitro studies) has been accumulated in the last 40 years, which attributes the light-stimulated wound-healing process and the immune response to fibroblast growth and locomotion,[73] increased cell proliferation, cell activation, cell division, cell maturation, the release of interleukins, mast cell degranulation, collagen synthesis,[74] secretion of growth factors, DNA synthesis, ATP production,[75] osteoblast proliferation,[76] and calcium uptake by macrophages.[77] The particular mechanism of phototherapy at the cellular level is not entirely understood; however, there is growing evidence that the primary mechanism is based on the absorption of monochromatic visible and NIR radiation by components of the cellular respiratory chain.[78] Mitochondria are thought to be a likely site for the initial effects of light, leading to increased ATP production, modulation of Reactive Oxygen Species, and induction of transcription factors. These effects, in turn, lead to increased cell proliferation and migration (particularly by fibroblasts), modulation in levels of cytokines, growth factors, and inflammatory mediators, and increased tissue oxygenation. In particular, Cytochrome c Oxidase (COX) is mammalian cells' primary photo acceptor between 630 nm and 900 nm,[79] responsible for more than 50% of absorption longer than 800 nm.

However, COX, as the primary mitochondrial photoacceptor for R-NIR photons, is not the only explanation. In particular, Sommer et al.[80] challenged this long-standing view. They looked at mitochondria as a Field-Effect Transistor (FET). The functional interplay of cytochrome c (emitter) and COX (drain) with a nanoscopic interfacial water layer (gate) between the two enzymes forms a biological FET in which R-NIR photons control the gate. By reducing the viscosity of the nanoscopic interfacial water layers within and around the mitochondrial rotary motor in oxidatively stressed cells, R-NIR light promotes the synthesis of extra Adenosine Triphosphate (ATP).

To date, laboratory investigations (cells and cultures) and experiments on animal models form the bulk of published research on the effectiveness of phototherapy for chronic wound healing.

In particular, Spitler and Berns[81] compared the efficacy of visible light at different wavelengths in promoting wound healing. The authors found laser light at 652 nm (10 J/cm^2) and 806 nm (2.3 J/cm^2), as well as LED at 637 nm (10.02 J/cm^2) and 901 nm (2.3 J/cm^2) induced comparable levels of cell migration and wound closure.

Various positive phototherapy effects (vasodilation, bone formation enhancement, and wound healing acceleration) have been reported on rodent models. For example, Dancakova et al.[82] demonstrated that infrared 810 nm laser (0.9 J/cm^2/wound/day) light improved wound healing in diabetic rats with respect to the untreated group. In a mouse model, Gupta et al.[83] demonstrated that 635 nm and 810 nm light, delivered at a constant fluence (4 J/cm^2), was effective in promoting healing in dermal abrasions. In contrast, 730 nm and 980 nm light showed no sign of stimulated healing.

Less positive results were found on pig models, which are more relevant to the human skin.

6.4.4 Open Questions for LLLT Therapy

Despite many reports of positive findings from experiments conducted in vitro and in animal models, randomized controlled clinical trials of phototherapy remain controversial. This likely is due to two main reasons. Firstly, the biochemical mechanisms are incompletely understood. In particular, a biphasic dose response has been frequently observed where low light levels have a much better effect than higher levels. It has been found that there exists an optimal dose of light for any particular application, and doses lower than this optimum value, or more significantly, larger than the optimum value, will have a diminished therapeutic outcome, or for high doses of light, a negative outcome may even result.

Secondly, it is extremely difficult to choose rationally amongst a large number of illumination parameters such as wavelength, fluence, power density, pulse structure, and treatment timing. So, further research in cell mechanisms and delivery methods is required.

Also, there is some related controversy around coherency and monochromaticity. Laser is a coherent and monochromatic source of light. Thus, initially, these features were thought to be primarily responsible for therapeutic effects. However, these assumptions have been questioned with the introduction of LEDs, which produce a noncoherent and quasi-monochromatic light. Nevertheless, monochromaticity (or quasi-monochromaticity) is considered pertinent to the use of light as a therapy as it has been shown that the effects present with narrow-band light are absent when broad-spectrum light is used.[84]

One of the most important limitations to advancing phototherapy into mainstream medical practice is the lack of appropriately controlled and blind clinical trials. However, the primary limitation is the absence of a valid photon-cell interaction model. As a result, progress in LLLT is realized by a trial-and-error process instead of a systematic approach based on established basic science.

NOTES

1. DG Armstrong, BA Lipsky, "Diabetic foot infections: Stepwise medical and surgical management", *Int Wound J*, 1(2):123–132 (2004).
2. Centers for Disease Control and Prevention, US Department of Health and Human Services, *National Diabetes Fact Sheet: General Information and National Estimates on Diabetes in the United States, 2003*, US Centers for Disease Control and Epidemiology, Atlanta, GA, 2003.
3. Centers for Disease Control and Prevention. National Diabetes Statistics Report website. https://www.cdc.gov/diabetes/data/statistics-report/index.html. Accessed Feb 17, 2023.
4. J Whitney, L Phillips, R Aslam, et al. "Guidelines for the treatment of pressure ulcers", *Wound Repair Regen*, 14(6): 663–679 (2006).
5. TS Alster, JR Lupton, "Erbium:YAG cutaneous laser resurfacing", *Dermatol Clin*, 19(3):453–466 (2001).
6. S Stellar, R Meijer, S Walia, S Mamoun, "Carbon dioxide laser debridement of decubitus ulcers: Followed by immediate rotation flap or skin graft closure", *Ann Surg*, 179(2):230–237 (1974).
7. D Evison, RF Brown, P Rice, "The treatment of sulphur mustard burns with laser debridement", *J Plast Reconstr Aesthet Surg*, 59(10):1087–1093 (2006).
8. N Reynolds, N Cawrse, T Burge, J Kenealy, "Debridement of a mixed partial and full thickness burn with an erbium:YAG laser", *Burns*, 29(2):183–188 (2003).
9. H Guan, D Zhang, X Ma, et al. "Efficacy and safety of CO2 laser in the treatment of chronic wounds: A retrospective matched cohort trial", *Lasers Surg Med*, 54: 490–501 (2022).
10. Efficacy of the Er:YAG Laser Debridement on Patient-Reported Pain and Bacterial Load in Chronic Wounds, ClinicalTrials.gov Identifier: NCT03182582.
11. LS Bass, MR Treat, "Laser tissue welding: A comprehensive review of current and future clinical applications", *Lasers Surg Med*, 17(4): 315–349 (1995).

12. MM Judy, JL Matthews, RL Boriak, et al. "Heat-free photochemical tissue welding with 1, 8-naphtalinide dye using visible (420nm) light," in *OE/LASE'93: Optics, Electro-Optics, & Laser Applications in Science & Engineering*, 1993, pp. 175–179.
13. L Mulroy, J Kim, I Wu, et al. "Photochemical keratodesmos for repair of lamellar corneal incisions", *Invest Ophthalmol Vis Sci*, 41(11): 3335–3340 (2000).
14. M Yao, A Yaroslavsky, FP Henry, et al. "Phototoxicity is not associated with photochemical tissue bonding of skin", *Lasers Surg Med*, 42(2): 123–131 (2010).
15. J Whitney, L Phillips, R Aslam, et al. "Guidelines for the treatment of pressure ulcers", *Wound Repair Regen*, 14(6): 663–679 (2006).
16. TA Conner-Kerr, PK Sullivan, J Gaillard, et al. "The effects of ultraviolet radiation on antibiotic-resistant bacteria in vitro", *Ostomy Wound Manage*, 44(10): 50–56 (1998).
17. TP Coohill, J-L Sagripanti, "Overview of the inactivation by 254 nm ultraviolet radiation of bacteria with particular relevance to biodefense", *Photochem Photobiol*, 84: 1084–1090 (2008).
18. M Buonanno, D Welch, I Shuryak, DJ Brenner, "Far-UVC light (222 nm) efficiently and safely inactivates airborne human coronaviruses," *Sci Rep*, 10:10285 (2020).
19. O Raab,"Uber die wirkung flureszierender stoffe auf infusorien", *Zeitschr Biol*,39:524–546 (1900).
20. JD Macmillan, WA Maxwell, C Chichester, "Lethal photosensitization of microorganisms with light from a continuous-wave gas laser", *Photochem Photobiol*, 5(7):555–565 (1966).
21. G Jori, SB Brown, "Photosensitized inactivation of microorganisms", *Photochem Photobiol Sci*, 3(5): 403–405 (2004).
22. MR Hamblin, DA O'Donnell, N Murthy, et al. "Rapid control of wound infections by targeted photodynamic therapy monitored by in vivo bioluminescence imaging", *Photochem Photobiol*, 75(1): 51–57 (2002).
23. S Morley, J Griffiths, G Philips, et al. "Phase IIa randomized, placebo-controlled study of antimicrobial photodynamic therapy in bacterially colonized, chronic leg ulcers and diabetic foot ulcers: A new approach to antimicrobial therapy", *Br J Dermatol*, 168(3): 617–624 (2013).
24. CS Foote, "Definition of type I and type II photosensitized oxidation", *Photochem Photobiol*, 54(5):659 (1991).
25. J Bellin, L Lutwick, B Jonas, "Effects of photodynamic action on E. coli", *Arch Biochem Biophys*,132(1):157–164 (1969).
26. J Ghorbani, D Rahban, S Aghamiri, et al. "Photosensitizers in antibacterial photodynamic therapy: An overview", *Laser Therapy*, 27(4): 293–302 (2018).
27. NS Soukos, S Som, AD Abernethy, et al. "Phototargeting oral black-pigmented bacteria", *Antimicrob Agents Chemother*, 49(4):1391–1396 (2005).
28. AM Lennon, W Buchalla, L Brune, et al. "The ability of selected oral microorganisms to emit red fluorescence", *Caries Research*, 40(1):2–5 (2006).

29. F Cieplik, A Spath, C Leibl, et al. "Blue light kills aggregatibacter actinomy- cetemcomitans due to its endogenous photosensitizers", *Clinical Oral Investigations*, 18(7):1763–1769 (2014).
30. O Simonetti, O Cirioni, F Orlando, et al. "Effectiveness of antimicrobial photodynamic therapy with a single treatment of RLP068/Cl in an experimental model of Staphylococcus aureus wound infection", *Brit J Dermat*, 164(5):987–995 (2011).
31. MQ Mesquita, JC Menezes, MG Neves, et al. "Photodynamic inactivation of bioluminescent Escherichia coli by neutral and cationic pyrrolidine-fused chlorins and isobacteriochlorins", *Bioorg &Med Chem Lett*, 24(3):808–812 (2014).
32. I Garcia, S Ballesta, Y Gilaberte, et al. "Antimicrobial photodynamic activity of hypericin against methicillin-susceptible and resistant Staphylococcus aureus biofilms", *Future Microbiology*, 10(3):347–356 (2015).
33. K Parvathy, P Negi, P Srinivas, "Antioxidant, antimutagenic and antibacterial activities of curcumin-ß-diglucoside", *Food Chem*, 115(1):265–271 (2009).
34. T Maisch, A Eichner, A Späth, et al. "Fast and effective photodynamic inactivation of multiresistant bacteria by cationic riboflavin derivatives", *PloS One*, 9(12):e111792 (2014).
35. S Perni, P Prokopovich, J Pratten, et al. "Nanoparticles: Their potential use in antibacterial photodynamic therapy", *Photochem & Photobiol Sci*, 10(5):712–720 (2011).
36. T Tsai, YT Yang, TH Wang, et al. "Improved photodynamic inactivation of gram-positive bacteria using hematoporphyrin encapsulated in liposomes and micelles", *Lasers Surg Med*, 41(4):316–322 (2009).
37. I Banerjee, D Mondal, J Martin, RS Kane, "Photoactivated antimicrobial activity of carbon nanotube- porphyrin conjugates", *Langmuir*, 26(22):17369–17374 (2010).
38. J Gil-Tomás, S Tubby, IP Parkin, et al. "Lethal photosensitisation of Staphylococcus aureus using a toluidine blue O–tiopronin–gold nanoparticle conjugate", *J Mater Chem*, 17(35):3739–3746 (2007).
39. GP Tegos, TN Demidova, D Arcila-Lopez, et al. "Cationic fullerenes are effective and selective antimicrobial photosensitizers", *Chem Biol*, 12(10):1127–1135 (2005).
40. K Nguyen, A Khachemoune, "An update on topical photodynamic therapy for clinical dermatologists", *J Dermat Treat*, 30(8): 732–744 (2019).
41. YJ Lee, YC Yi, YC Lin, et al. "Purification and biofabrication of 5-aminolevulinic acid for photodynamic therapy against pathogens and cancer cells", *Bioresour Bioprocess*, 9: 68 (2022).
42. DH Hawkins-Evans, H Abrahamse, "Efficacy of three different laser wavelengths for in vitro wound healing",*Photodermatol Photoimmunol Photomed*, 24: 199–210 (2008).
43. E Bornstein, W Hermans, S Gridley, et al. "Near-infrared photoinactivation of bacteria and fungi at physiologic temperatures", *Photochem Photobiol*, 85: 1364–1374 (2009).

44. SL Percival, I Francolini, G Donelli, "Low-level laser therapy as an antimicrobial and antibiofilm technology and its relevance to wound healing", *Future Microbiology*, 10(2): 255–272 (2015).
45. P Papageorgiou, A Katsambas, A Chu, "Phototherapy with blue (415 nm) and red (660 nm) light in the treatment of acne vulgaris", *Br J Dermatol*, 142: 973–978 (2000).
46. M Maclean, SJ MacGregor, JG Anderson et al. "Inactivation of bacterial pathogens following exposure to light from a 405-nanometer light-emitting diode array", *Appl Environ Microbiol*, 75: 1932–1937 (2009).
47. M Denzinger, M Held, S Krauss, et al. "Does phototherapy promote wound healing? Limitations of blue light irradiation", *Wounds*, 33(4):91–98 (2021).
48. PE Hockberger, TA Skimina, VE Centonze et al. "Activation of flavin-containing oxidases underlies light-induced production of H_2O_2 in mammalian cells", *Proc Natl Acad Sci USA*, 96(11): 6255–6260 (1999).
49. MR Hamblin, T Hasan, "Photodynamic therapy: A new antimicrobial approach to infectious disease?" *Photochem Photobiol Sci*, 3: 436–450 (2004).
50. M Wainwright, "Photodynamic antimicrobial chemotherapy (PACT)", *J Antimicrob Chemother*, 42: 13–28 (1998).
51. M Maclean, SJ MacGregor, JG Anderson et al. "Inactivation of bacterial pathogens following exposure to light from a 405-nanometer light-emitting diode array", *Appl Environ Microbiol*, 75: 1932–1937 (2009).
52. CC Riddle, SN Terrell, MB Menser, et al. "A review of photodynamic therapy (PDT) for the treatment of acne vulgaris", *J Drugs Dermatol*, 8(11):1010–1019 (2009).
53. B Fournier, DJ Philpott, "Recognition of Staphylococcus aureus by the innate immune system", *Clin Microbiol Rev*, 18: 521–540 (2005).
54. SL Percival, I Francolini, G Donelli, "Low-level laser therapy as an antimicrobial and antibiofilm technology and its relevance to wound healing", *Future Microbiology*, 10(2):255–272 (2015).
55. AH Coulter, "Let there be light-and healing", *Altern Complement Ther*, 8(6): 322–326 (2003).
56. J Basford, "Low intensity laser therapy: Still not an established clinical tool", *Lasers Surg Med*, 16:331–342 (1995).
57. T Moshkovska, J Mayberry, "It is time to test low level laser therapy in Great Britain", *Postgrad Med J*, 81(957):436–441 (2005).
58. LA Santana-Blank, E Rodríguez-Santana, KE Santana-Rodríguez, "Photo-infrared pulsed bio-modulation (PIPBM): A novel mechanism for the enhancement of physiologically reparative responses", *Photomed Laser Surg*, 23(4): 416–424 (2005).
59. ML Kripke, "Ultraviolet radiation and immunology: Something new under the sun – presidential address", *Cancer Res*, 54(23): 6102–6105 (1994).
60. P Whittaker, "Laser acupuncture: Past, present, and future", *Lasers Med Sci*, 19(2): 69–80 (2004).
61. AY Douplik, YN Gordeev, IV Yaroslavsky, "Calculation of specific power density of blood for intravenous low-level laser therapy," in *Radiofrequency and Optical Methods of Biomedical Diagnostics and Therapy*, 1993, pp. 240–244.

62. E Mester, G Ludany, M Selyei, et al. "The stimulating effect of low power laser rays on biological systems", N. p., 1968 Web.
63. CS Enwemeka, "Light is light", *Photomed Laser Surg*, 23(2): 159–160 (2005).
64. A Schindl, M Schindl, H Schon, et al. "Low-intensity laser irradiation improves skin circulation in patients with diabetic microangiopathy", *Diabetes Care*, 21(4):580–584 (1998).
65. N Kazemi-Khoo, "Successful treatment of diabetic foot ulcers with low-level laser therapy", *The Foot*, 16:184–187 (2006).
66. BM Kajagar, AS Godhi, A Pandi et al. "Efficacy of low level laser therapy on wound healing in patients with chronic diabetic foot ulcers-a randomised control trial", *Indian J Surg*, 74(5): 359–363 (2012).
67. EI Nussbaum, H Flett, SL Hitzig et al. "Ultraviolet-C irradiation in the management of pressure ulcers in people with spinal cord injury: A randomized, placebo-controlled trial", *Arch Phys Med Rehabil*, 94: 650–659 (2013).
68. C Lucas, MJ van Gemert, RJ de Haan, "Efficacy of low-level laser therapy in the management of stage III decubitus ulcers: A prospective, observer-blinded multicentre randomised clinical trial", *Lasers Med Sci*, 18(2): 72–77 (2003).
69. V Schubert, "Effects of phototherapy on pressure ulcer healing in elderly patients after a falling trauma. A prospective, randomized, controlled study", *Photodermatol Photoimmunol Photomed*, 17: 32–38 (2001).
70. AL Whinfield, I Aitkenhead, "The light revival: Does phototherapy promote wound healing? A review", *Foot (Edinb)*, 19(2):117–124 (2009).
71. EL Nussbaum, I Biemann, B Mustard, "Comparison of ultrasound/ultraviolet-C and laser for treatment of pressure ulcers in patients with spinal cord injury", *Phys Ther*, 74: 812–823 (1994).
72. Z Landau, A Sommer, EB Miller, "Topical hyperbaric oxygen and low-energy laser for the treatment of chronic ulcers", *Eur J Intern Med*, 17(4): 272–275 (2006).
73. D Hawkins, H Abrahamse, "Effect of multiple exposures of low-level laser therapy on the cellular responses of wounded human skin fibroblasts", *Photomed Laser Surg*, 24(6): 705–714 (2006).
74. AN Pereira, C de Paula Eduardo, E Matson, MM Marques, "Effect of low-power laser irradiation on cell growth and procollagen synthesis of cultured fibroblasts", *Lasers Surg Med*, 31(4): 263–267 (2002).
75. J Kujawa, L Zavodnik, I Zavodnik, et al. "Effect of low-intensity (3.75-25 J/cm2) near-infrared (810 nm) laser radiation on red blood cell ATPase activities and membrane structure", *J Clin Laser Med Surg*, 22(2): 111–117 (2004).
76. A Stein, D Benayahu, L Maltz, U Oron, "Low-level laser irradiation promotes proliferation and differentiation of human osteoblasts in vitro", *Photomed Laser Surg*, 23(2): 161–166 (2005).
77. S Young, M Dyson, P Bolton, "Effect of light on calcium uptake by macrophages," in Fourth International Biotherapy Association Seminar on Laser Biomodulation, 1991.

78. T Karu, "Laser biostimulation: A photobiological phenomenon", *J Photochem Photobiol B*, 3(4): 638–640 (1989).
79. TI Karu, NI Afanas'eva, "Cytochrome c oxidase as the primary photoacceptor upon laser exposure of cultured cells to visible and near IR-range light", *Dokl Akad Nauk,* 342(5): 693–695 (1995).
80. AP Sommer, P Schemmer, AE Pavláth, et al. "Quantum biology in low level light therapy: Death of a dogma", *Ann Transl Med*, 8(7): 440 (2020).
81. R Spitler, MW Berns, "Comparison of laser and diode sources for acceleration of in vitro wound healing by low-level light therapy", *J Biomed Opt*, 19: 38001 (2014).
82. L Dancakova, T Vasilenko, I Kovac, et al. "Low-level laser therapy with 810 nm wavelength improves skin wound healing in rats with streptozotocin-induced diabetes", *Photomed Laser Surg*,32(4):198–204 (2014).
83. A Gupta, T Dai, MR Hamblin, "Effect of red and near-infrared wavelengths on low-level laser (light) therapy-induced healing of partial-thickness dermal abrasion in mice", *Lasers Med Sci*, 29: 257–265 (2014).
84. JR Basford, "Low intensity laser therapy: Still not an established clinical tool", *Lasers Surg Med*, 16(4): 331–342 (1995).

CHAPTER 7

Future Directions

THROUGHOUT THIS BOOK, WE have debated the state of the art of optical technologies for managing diabetic ulcers. Several trends in medicine will likely drive the future development of optical wound care modalities. Therefore, in this chapter, we will briefly discuss them.

7.1 ARTIFICIAL INTELLIGENCE

With the rapid penetration of Artificial Intelligence (AI) into all aspects of human life, including the medical field, it is expected that it will be instrumental in the smart analysis of data derived from different sensors and relevant technologies for screening, diagnosis, and care of diabetic foot.

In particular, a review by Chemello et al.[1] identified a rapid influx of AI-based studies, which accelerated since 2020. The authors stratified approaches into several groups.

The first group of approaches aims at using AI-based techniques to automatically classify the risk of developing diabetic foot syndrome before any visual change can be perceived. These methods typically do not require clinical examinations, physical contact with the patients, or foot imaging. They are typically based on questionnaires,[2] genomic DNA,[3] or analysis of patients' health and socioeconomic data.[4]

The second group of approaches tries to predict the risk of ulceration based on imaging technologies. They are typically based on traditional RGB (anatomical, in our classification)[5] or thermographic[6] imaging. However, some groups started experimenting with hyperspectral imaging as well.[7]

The third group refers to studies of existing wounds without explicitly exploiting imaging techniques. For this group of studies, some quantitative measures (like wound area, duration, depth, site, arterial flow, BMI, and history of dialysis),[8] wound classification,[9] biochemical markers,[10] or in vitro bacteria samples[11] are used.

Finally, the fourth group of studies is based on imaging technologies applied to existing wounds. It combines both anatomical and physiological imaging. Most of these studies exploit AI-assisted image-based wound measurements,[12] wound localization,[13] and classification.[14] However, some studies exploit infection and ischemia detection.[15]

Integration of sensors and imaging modalities (both anatomic and physiologic) with AI for diabetic foot monitoring was also discussed by Kaselimi et al.[16]

While many AI-based approaches have been tried, the clinical adoption of AI-based technologies is still in its infancy. In addition to technology immaturity and small training sets, the primary obstacle right now is regulatory clearance. Like any other medical software, AI tools are medical devices regulated by market regulators like the US Food and Drug Administration (FDA). So far, regulators have taken a very cautionary approach to the clearance/approval of AI-based medical devices. For example, the software development lifecycle for AI-based tools differs significantly from well-defined traditional medical software development. In particular, AI models aim to learn from new data. However, it contradicts traditional regulatory approaches where any changes to the medical device need to be properly managed. As such, regulators actively monitor the field and strive to develop a framework that will help streamline AI-based medical device development. For example, in October 2021, the US Food and Drug Administration (FDA), Health Canada, and the United Kingdom's Medicines and Healthcare Products Regulatory Agency (MHRA) jointly identified ten guiding principles that can inform the development of Good Machine Learning Practice (GMLP).[17]

7.2 THERANOSTICS AND NANOMEDICINE

New materials, including nanomaterials, have significantly opened up the field of theranostics, which refers to a field of medicine that combines diagnostics and therapeutics into a single integrated approach. The term 'theranostics' is a combination of 'therapy' and 'diagnostics,' reflecting the goal of simultaneously diagnosing and treating diseases.

Traditionally, diagnostics and therapeutics have been separate stages in the medical process. Diagnostics involve identifying and characterizing diseases or conditions, while therapeutics involve treating or managing those diseases. However, theranostics aims to merge these two processes to create a more personalized and precise approach to patient care.

While being a buzzword, theranostics has yet to demonstrate its clinical utility. As a relatively new concept, it is established just in several medical fields, particularly in nuclear medicine, where it couples diagnostic imaging and therapy using the same molecule or at least very similar molecules, which are either radiolabeled differently or given in different dosages.[18]

However, it has significant potential in other medical fields, including wound care. For example, one can think of nanoparticles, which can be used for both diagnostics/visualization and treatment of wounds or certain skin conditions. A prototype here can be porphyrins produced by certain bacteria, which can be used for both bacteria localization (at small light intensities) and as an antimicrobial photosensitizer (at higher light intensities) (see Section 6.2.2.2 for more details).

New materials, like quantum dots (QDs) represent a promising direction in nanomedicine. QDs, – also called semiconductor nanocrystals, are semiconductor particles a few nanometers in size, having optical and electronic properties that differ from those of larger particles as a result of quantum mechanics. They are a central topic in nanotechnology and materials science. When the quantum dots are illuminated by UV light, an electron in the quantum dot can be excited to a state of higher energy. In the case of a semiconducting quantum dot, this process corresponds to the transition of an electron from the valence band to the conductance band.

QDs have unique properties that offer great utility to fluorescence imaging.[19] (1) specificity – the synthesis of QDs results in organic capping ligands that make them biocompatible, and suitable for biological targeting development. This is achieved by surface modification and linking with antibodies, peptides or small molecules. In effect, the QDs are modified to become specific ligands which discretely couple with their target. (2) Adjustable emission – narrow emission band and size-tunable Gaussian emission spectra in the visible spectrum. (3) Strong signals – QDs have a high fluorescence quantum yield and a large Stokes shift.

Thus, it is reasonable to expect that with nanomedicine developments, we will see many new applications, including theranostic ones.

7.3 WIDE USE OF SMARTPHONES

Smartphones have the potential to revolutionize wound care by enabling convenient and accessible methods for wound assessment, monitoring, and communication between healthcare providers and patients. Nowadays, they are powerful edge-computing devices supplied with high-resolution cameras. Some of them are already equipped with extra modalities, like NIR or thermal imaging. As such, they have the potential to bring down healthcare costs significantly. Here are some ways in which smartphones are being utilized in wound care:

1. Anatomical Wound Imaging: smartphones equipped with high-resolution cameras can capture detailed images of wounds. These images can be securely transmitted to healthcare providers for remote assessment, consultation, or documentation. This helps track wound progression over time and facilitates communication between healthcare professionals.

2. Wound Measurement Apps: there are smartphone apps that utilize built-in cameras and special measurement algorithms to accurately measure wound dimensions, including length, width, and area. This can assist in documenting wound size and tracking healing progress without requiring specialized equipment.

3. Physiological Wound Imaging: smartphones can be paired with other devices (like dermascope or otoscope) to create sophisticated medical devices. They are equipped with high-resolution cameras and powerful processors, enabling multiple modalities, including HSI/MSI, fluorescence imaging, etc.

4. Telemedicine and Virtual Consultations: with the rise of Telemedicine (TM), smartphones can be used to conduct virtual consultations with healthcare providers. Patients can share images or videos of their wounds in real-time, allowing healthcare professionals to assess and provide recommendations remotely.

5. Wound Care Management Apps: there are smartphone applications designed specifically for wound care management. These apps may include features such as reminders for dressing changes, medication schedules, wound care instructions, and tracking wound-healing progress. They can help patients and caregivers stay organized and adhere to their wound care plan.

6. Patient Education and Self-management: smartphone apps can provide educational resources on wound care, including videos, diagrams, and step-by-step instructions. Patients can access these resources at their convenience to learn about proper wound care techniques, prevention of complications, and self-management strategies.

7. Remote Monitoring and Alerts: smartphones can be connected to wearable sensors or smart dressings that monitor parameters like temperature, moisture levels, or pH at the wound site. This data can be transmitted to healthcare providers in real-time, enabling early detection of potential issues or complications.

However, it is important to note that while smartphones offer many benefits in wound care, maintaining patient privacy and data security is crucial. Developers and healthcare providers must ensure adherence to relevant regulations and guidelines to protect patient information when utilizing smartphones in wound care.

7.4 MULTIMODAL DEVICES

The traditional approach in medical device development is to develop a specific device for a specific (often narrow) task. However, it led to ORs, ICUs, or labs being clogged with multiple devices. Moreover, these devices are often incompatible or interoperable. A multimodal approach in medical imaging refers to using multiple imaging modalities, such as combining different imaging techniques or modalities to provide a more comprehensive and accurate assessment of a patient's condition. Combining several modalities into a single device is the natural approach to multimodality. A multimodal approach has significant benefits, particularly in the context of image registration, synchronization, and increased specificity:

1. Improved Visualization and Understanding: by combining information from multiple imaging modalities, a multimodal approach allows for a more complete visualization of anatomical structures, functional processes, and tissue characteristics. This comprehensive visualization helps to understand better the relationship between different structures and can reveal crucial diagnostic information that may be missed using a single modality alone.

2. Enhanced Image Registration: image registration aligns images from different modalities or time points. By aligning images accurately, it becomes easier to analyze changes over time, compare images side by side, and precisely locate and track specific features or abnormalities in the patient's anatomy. However, typically, it represents a significant technical challenge. The multimodal approach in a single device solves an image registration problem.

3. Synchronization of Information: synchronization of information from several devices represents another significant technical challenge. Multimodal devices solve a synchronization problem.

4. Increased Specificity and Diagnostic Accuracy: each imaging modality has strengths and limitations in visualizing different aspects of tissue characteristics or pathology. By combining multiple modalities, a multimodal approach can enhance imaging findings' specificity and diagnostic accuracy. For example, consider the clinical question, 'Is the wound infected?' Assume two imaging modalities. The first is infrared thermography which indicates that the wound is infected and has a positive predictive value of 90%. The second is bacterial fluorescence which has a positive predictive value of 80%. If both are positive, the predictive value for infection becomes 98%.

5. Personalized Treatment Planning: a multimodal approach in imaging enables a more personalized treatment planning process. By integrating information from various modalities, clinicians can understand better the patient's condition and tailor treatment plans to their specific needs. This can lead to more effective and targeted interventions, such as surgical guidance or monitoring response to treatment.

6. Saving Space and Money: multimodal devices may significantly impact unclogging ORs, ICUs, and labs while decreasing overall costs.

Overall, a multimodal approach in medical imaging offers advantages in terms of improved visualization, enhanced image registration, synchronized information, increased specificity, and personalized treatment planning. By leveraging the strengths of different imaging modalities, healthcare professionals can obtain a more comprehensive assessment and make informed clinical decisions for optimal patient care.

7.5 INTEROPERABILITY: INTEGRATION WITH EHR SYSTEMS

Many medical devices are still standalone devices. However, the ability to integrate the device into the Electronic Health Record (EHR) system is becoming an increasingly important feature. There are several benefits of integration:

1. Comprehensive Patient Information: integration allows for the seamless transfer of wound care data and information between wound care applications or devices and the EHR system. This ensures that wound-related information is integrated with the patient's overall medical record, providing a comprehensive view of their health status. It enables healthcare providers to access complete and up-to-date information about the patient, including wound history, treatment plans, and outcomes.

2. Care Coordination and Continuity: integration with EHR systems supports better care coordination among healthcare providers involved in the patient's treatment. It allows different healthcare professionals, such as wound care specialists, primary care physicians, nurses, and specialists, to access and contribute to the patient's wound care data in a coordinated manner. This promotes continuity of care, reduces the risk of duplicate or conflicting treatments, and facilitates collaboration among the care team.

3. Improved Clinical Decision-making: integration with EHR systems gives healthcare providers a more holistic and accurate picture of the patient's health. Access to complete wound care data and other clinical information helps make informed decisions about treatment plans, wound assessment, and interventions. As noted earlier, AI is largely based on weighted multivariate analysis. Integration into the medical record paves the way to metadata access and sophisticated risk / success stratification. For example, what is the probability of healing for wound that has closed by 35% over a four-week period in a 40-year-old patient vs. a 70-year-old patient. To this equation one could layer numerous additional risk factors such as cardiac ejection fraction, hemoglobin A1c (blood glucose control), renal function, hemoglobin levels, and many more metadata points. By embedding the imaging analytics into the EHR, the extraction and weighting of numerous data points would guide important decisions such

as amputation vs. continued treatment. EHR integration supports evidence-based care by triggering clinical alerts for consideration of therapeutic or diagnostic options which are frequently overlooked.[20]

4. Efficiency and Time Savings: integration reduces the need for manual data entry and duplication of efforts. Wound care data captured through mobile devices or specialized wound care applications can be automatically transferred to the EHR system. This eliminates the need for healthcare providers to spend time entering data manually, enabling them to focus more on patient care and reducing the potential for errors in data entry.

5. Research, Advanced Algorithms, and Reporting: integrated wound care data in EHR systems can be used for research, training of advanced algorithms, quality improvement initiatives, and reporting purposes. Aggregated and anonymized wound care data from multiple patients can help identify trends, evaluate the effectiveness of treatment strategies, and contribute to the advancement of wound care practices.

Thus, integration with EHR systems ensures that wound care data are accessible, shared, and utilized effectively, leading to better care coordination, improved decision-making, and enhanced patient outcomes.

There are several levels of integration with EHR systems: no integration, generic integration, and full integration.

No integration refers to the case where the captured information is either not stored or stored locally on the device. In the latter case, the information can be transferred manually to the computer and entered into the EHR system if necessary.

With virtually thousands of available EHR systems, it is becoming increasingly difficult to integrate them or to communicate between them. Thus, generic interoperability protocols like HL7 or, more recently, Fast Healthcare Interoperability Resources (FHIR) have been developed. They provide basic interoperability between various EHR systems and can be used for generic medical device integration.

The third level of integration is the full integration with a particular EHR system. It gives significant advantages over no integration or generic integration, particularly in functionality. However, how is it possible to integrate a medical device with a multitude of existing EHR systems? The answer is that there is probably no need to do it. If the EHR market was fragmented 10 years ago, then by the time of writing, the market has become much less fragmented, with clear winners emerging in the so-called verticals (e.g. hospitals or home healthcare). For example, Epic (Epic Systems, Wisconsin, US) is a de facto EHR system for hospitals in the US. Thus, for any given target market (or vertical), certain EHR systems can be targeted first to get the most coverage.

7.6 TELEMEDICINE AND REMOTE MONITORING: HEALTHCARE CLOSER TO PATIENT

Oftentimes, patients present to healthcare providers at later stages of the disease, when more serious interventions are required. If wound care can be moved closer to the patient, the disease can be detected earlier, thus improving outcomes and the quality of the patient's life. Optical modalities can be integrated into Telemedicine and Remote Monitoring, thus bringing diagnostics and disease management closer to the patient. For example, telemedicine technologies can remotely monitor wounds, provide consultation, and facilitate follow-up care, reducing the need for frequent in-person visits.

In particular, diabetic ulcers are traditionally managed through hospital or clinic appointments with multiple specialists such as wound care physicians, podiatrists, infectious disease specialists, vascular surgeons, and wound care nurses. Diabetic ulcers are typically monitored weekly for several months before they heal completely. These ulcers often require consultations with a range of specialists. Predictably, outcomes for DFUs are substantially worse for patients residing in remote areas with limited access to specialists. In rural areas, amputations are up to ten times higher than in urban areas.[21]

Thus, telehealth has great appeal for patients in rural areas and those with transport barriers in urban areas.

The Telemedicine (TM) solution helps save time by reducing the frequency of visits to wound clinics. This was reported from Australian and Canadian sources where patients living in distant rural areas had to take several days off to attend a consultation. Furthermore, patients living in rural areas also reported the benefit of avoiding stress related to driving in large unfamiliar cities, looking for parking spaces, eating places, and toilet facilities.[22]

Patient-focused studies have found that patients did not want their relatives or friends to be obliged to help. Patients would rather have an equal and normal relationship with their family and friends. The TM solutions increased their experience of independence, contributing to decreased guilt of being a burden.[23] Also, expenses related to the workplace, where they had to miss a day's work, often with the result that they missed a day's pay.[24]

In many countries, the use of different modalities of TM is already implemented as a communication tool between caregivers across the healthcare sectors. In the TM modalities, where patients were encouraged to report wound assessments using pictures, they experienced a greater feeling of responsibility and awareness for signs of infection.[25]

There is significant telemedicine interest among patients, providers, and payers as a useful vehicle to supplement traditional face-to-face care. However, the cost-effectiveness of DFU telehealth management is not fully understood. Published financial studies are fragmented and narrow in scope and only explore economics from a singular perspective, e.g., the provider, patient, hospital, etc. To date, no 'total cost accounting' study has considered all stakeholders over a long period of time.

In summary, while it is true that not all wound care can or should be delivered virtually, it is equally true that not all wound care must occur in the clinic.

NOTES

1. G Chemello, B Salvatori, M Morettini, A Tura, "Artificial intelligence methodologies applied to technologies for screening, diagnosis and care of the diabetic foot: A narrative review", *Biosensors* 12: 985 (2022).
2. ACBH Ferreira, DD Ferreira, HC Oliveira, et al. "Competitive neural layer-based method to identify people with high risk for diabetic foot", *Comput Biol Med*, 120: 103744 (2020).
3. K Singh, VK Singh, NK Agrawal, et al. "Association of Toll-like receptor 4 polymorphisms with diabetic foot ulcers and application of artificial neural network in DFU risk assessment in type 2 diabetes patients", *BioMed Res Int*, 2013: 318686 (2013).
4. Z Schäfer, A Mathisen, K Svendsen, et al. "Toward machine-learning-based decision support in diabetes care: A risk stratification study on diabetic foot ulcer and amputation", *Front Med*, 7: 601602 (2020).
5. CL Toledo Peral, FJ Ramos Becerril, G Vega Martínez, et al. "An application for skin macules characterization based on a 3-stage image-processing algorithm for patients with diabetes", *J Healthc Eng*, 2018: 9397105 (2018).
6. I Cruz-Vega, D Hernandez-Contreras, H Peregrina-Barreto, et al. "Deep learning classification for diabetic foot thermograms", *Sensors*, 20: 1762 (2020).
7. V Dremin, Z Marcinkevics, E Zherebtsov, et al. "A. Skin complications of diabetes mellitus revealed by polarized hyperspectral imaging and machine learning", *IEEE Trans Med Imaging*, 40: 1207–1216 (2021).

8. DJ Margolis, N Mitra, DS Malay, et al. "Further evidence that wound size and duration are strong prognostic markers of diabetic foot ulcer healing", *Wound Repair Regen*, 30: 487–490 (2022).
9. P Xie, Y Li, B Deng, et al. "An explainable machine learning model for predicting in-hospital amputation rate of patients with diabetic foot ulcer", *Int Wound J*, 19: 910–918 (2022).
10. C Lin, Y Yuan, L Ji, et al. "The amputation and survival of patients with diabetic foot based on establishment of prediction model", *Saudi J Biol Sci*, 27: 853–858 (2020).
11. N Yusuf, A Zakaria, MI Omar, et al. "In-vitro diagnosis of single and poly microbial species targeted for diabetic foot infection using e-nose technology", *BMC Bioinform*, 16: 158 (2015).
12. L Wang, PC Pedersen, E Agu, et al. "Area determination of diabetic foot ulcer images using a cascaded two-stage SVM-based classification", *IEEE Trans Biomed Eng*, 64: 2098–2109 (2017).
13. M Goyal, ND Reeves, S Rajbhandari, MH Yap, "Robust methods for real-time diabetic foot ulcer detection and localization on mobile devices", *IEEE J Biomed Health Inform*, 23: 1730–1741 (2019).
14. M Goyal, ND Reeves, AK Davison, et al. "DFUNet: Convolutional neural networks for diabetic foot ulcer classification", *IEEE Trans Emerg Top Comput Intell*, 4: 728–739 (2020).
15. M Goyal, ND Reeves, S Rajbhandari, et al. "Recognition of ischaemia and infection in diabetic foot ulcers: Dataset and techniques", *Comput Biol Med*, 117: 103616 (2020).
16. M Kaselimi, E Protopapadakis, A Doulamis, N Doulamis, "A review of non-invasive sensors and artificial intelligence models for diabetic foot monitoring", *Front Physiol*, 13:924546(2022).
17. https://www.fda.gov/medical-devices/software-medical-device-samd/good-machine-learning-practice-medical-device-development-guiding-principles (accessed Jun 25, 2023).
18. A Yordanova, E Eppard, S Kürpig, et al. "Theranostics in nuclear medicine practice", *Onco Targets Ther*, 10:4821–4828 (2017).
19. A Salleh, MB Fauzi, "The In Vivo, In Vitro and In Ovo Evaluation of Quantum Dots in Wound Healing: A Review". *Polymers*, 13: 191 (2021).
20. A Gawande, *The Checklist Manifesto: How to Get Things Right*, 1st Ed, Picador, London, 2011.
21. A Drovandi, S Wong, L Seng, et al. "Remotely delivered monitoring and management of diabetes-related foot disease: An overview of systematic reviews", *J Diab Scie Techn*, 17(1): 59–69 (2023).
22. SF Søndergaard, EG Vestergaard, AB Andersen, et al. "How patients with diabetic foot ulcers experience telemedicine solutions: A scoping review", *Int Wound J*, 20:1796–1810 (2023).
23. C Boodoo, JA Perry, PJ Hunter, et al. "Views of patients on using mHealth to monitor and prevent diabetic foot ulcers: Qualitative study", *JMIR Diabetes*, 2(2):e22–e (2017).

24. J Devine, "User satisfaction and experience with a telemedicine service for diabetic foot disease in an Australian rural community", Greater Southern Area Health Service (GSAHS) online Bega Valley Community Health. Greater Southern Area Health Service (GSAHS), Hujenni, 2007.
25. B Ploderer, R Brown, LSD Seng, et al. "Promoting self-care of diabetic foot ulcers through a mobile phone app: User-centered design and evaluation", *JMIR Diabetes*, 3(4):e10105–e (2018).

APPENDIX A
Skin and Wound Morphology

THE CHAPTER'S PRIMARY GOAL is to provide an overview of the structural and optical properties of the skin, which are important in medical practice, particularly for diagnostics. In addition, we provide an overview of wound morphology.

A.1 SKIN TYPES

The skin is the human body's largest organ, with a total area of approximately 2 sq. m. It plays a vital role in protection and thermoregulation. The skin can be subdivided into glabrous (non-hairy) and non-glabrous (hairy) skin. Glabrous (or hairless) skin is found on the palms of the hands, soles of the feet, and fingertips. As the primary point of contact with the external environment, glabrous skin differs from other skin types. In particular, it is much thicker and more durable. Additionally, glabrous skin is a major organ for sensing the external environment (somatosensation). As such, it is more sensitive to touch due to the high density of sensory receptors in this area.

Non-glabrous or hairy skin is found on the rest of the body. It is the prevalent skin type covering approximately 90% of the body surface. The presence of hair follicles characterizes non-glabrous skin.

The epidermis is less than 0.1 millimeters thick in hairy skin, and the dermis is 1–2 millimeters deep. Glabrous skin is thicker than hairy skin;

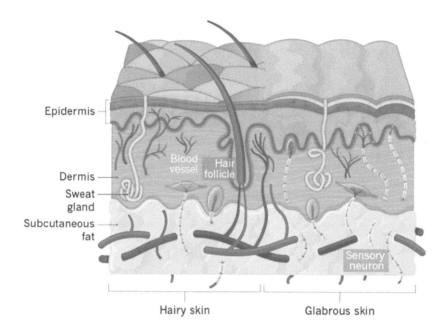

FIGURE A.1 Skin morphology: non-glabrous and glabrous skin. (Reproduced from[a] with permissions). J Gould, "Superpowered skin", *Nature*, 563(7732): S84 (2018).

the epidermis is about 1.5 millimeters thick, and the dermis is about 3 millimeters deep.

A.2 SKIN MORPHOLOGY

The skin is subdivided into three major layers: epidermis, dermis, and subcutaneous tissue. The composition of glabrous and non-glabrous skin is depicted in Figure A.1.

A.2.1 Epidermis

The epidermis is the outermost layer of the skin, and it plays a crucial role in protecting the body from external harm such as UV radiation and physical trauma. The epidermis comprises two sublayers, namely the non-living and living epidermis. It consists of four types of cells, namely keratinocytes (which produce keratin), melanocytes (which produce melanin), Langerhans cells, and Merkel cells. Among these cells, keratinocytes are the most abundant. They are arranged in five distinct layers (stratums), namely the stratum corneum, stratum lucidum, stratum granulosum, stratum spinosum, and stratum basale, each with a specific function.

The outermost layer of the epidermis is referred to as the stratum corneum. This layer comprises flattened, dead skin cells that have migrated to the skin's surface. This layer acts as a barrier to water loss and helps to protect the body from external insults.

Below the stratum corneum is the stratum lucidum, found only in glabrous skin. It is also composed of dead cells and packed with lipid-rich eleidin, which helps to keep water out. The stratum corneum and stratum lucidum form the non-living epidermis. The non-living epidermis is composed entirely of dead squamous cells that are highly keratinized and contain high amounts of lipids (~20%) and proteins (~60%).[1] This layer is usually around 20 micrometers thick and has relatively low water content, comprising only about 20% of its composition.

The living epidermis (~100 μm thick) consists of living cells produced in continually dividing basal cells, which push older cells upwards.

Below the non-living epidermis is the stratum granulosum, which is composed of keratinocytes that are in the process of dying and becoming flattened. Keratinocytes produce keratin as they mature. Keratin helps to strengthen the skin and protect it from damage.

The stratum spinosum is the next layer of the epidermis, composed of actively dividing cells. This layer primarily consists of keratinocytes held together by desmosomes.

The basal cell layer, also known as the stratum basale, is the deepest layer of the epidermis. It is composed of stem cells responsible for the continuous production of new skin cells. As these cells mature and move toward the skin's surface, they differentiate into the various cell types found in the other epidermis layers. A basal lamina separates the epidermis from the dermis.

In addition to the production of keratinocytes, the stratum basale contains melanin, which is responsible for most skin pigmentation and protects skin from the harmful effects of UV radiation. Melanin is produced in melanocytes and is found in membrane-bound intracellular organelles called melanosomes. There are two types of melanin: phaeomelanin, which appears reddish-yellow, and eumelanin, which appears brownish-black. The amount of melanin in the skin, which depends on the number of melanosomes present per unit volume of the epidermis, affects its ability to absorb light. In particular, the proportion of the epidermis occupied by melanosomes varies, ranging from 1.3% in individuals with light skin pigmentation to 43% in those with dark skin pigmentation.[2]

A.2.2 Dermis

The dermis is the inner layer of the skin, located below the epidermis. It is a thick and resilient layer of tissue that provides structural support to the skin. In addition, it provides sensory function and immune protection.

The dermis comprises two main layers: the papillary and reticular layers. The papillary layer is the dermis's outer layer, composed of loose, connective tissue that contains small blood vessels, nerve endings, and immune cells. This layer is responsible for providing nutrients and oxygen to the cells of the epidermis, as well as for detecting sensations such as touch and temperature.

The deeper layer of the dermis, known as the reticular layer, is primarily composed of dense connective tissue that contains collagen and elastin fibers. These fibers give the dermis strength and elasticity, allowing it to stretch and move without tearing. The reticular layer also contains larger blood vessels, sweat glands, and sebaceous glands, which are responsible for maintaining the body's temperature and moisture levels.

The dermis also contains various specialized cells, such as fibroblasts, which produce collagen and elastin fibers, and macrophages, which help defend the body against infection and disease. The dermis is a layer of the skin that contains a rich network of blood vessels, which gives it its characteristic pinkish color. The main absorbers in the visible spectral range in this layer are hemoglobin, carotene, and bilirubin. In the near-infrared spectral range, the absorption properties of the dermis are mainly defined by the amount of water present in it. The dermis also has a fibrous structure, mainly composed of collagen fibers that give it strength and elasticity. These fibers scatter light differently, depending on their size and arrangement. The papillary layer has smaller collagen fibers that scatter light in multiple directions, while the reticular layer has larger collagen fibers that scatter light mainly in a forward direction. The amount of blood and water also varies between the two layers, with the papillary layer having a higher blood volume fraction and the reticular layer having a higher water content. On average, the volume fraction of water in the dermis is estimated to be around 70%.[3] Blood content can be assessed from 0.2%[3] to 4%.[4]

A.2.3 Subcutaneous tissue

The subcutaneous tissue, also known as the hypodermis or subcutis, is the layer of tissue located beneath the dermis and above the underlying muscle and bone. It is a layer of fat and connective tissue that serves several critical bodily functions, including insulation, energy storage, and protection of the underlying tissues.

One of the main functions of the subcutaneous tissue is to provide insulation and energy storage. The fat cells in this layer, called adipocytes, store excess energy in triglycerides, which can be mobilized when the body needs additional fuel. The subcutaneous tissue also helps regulate body temperature by insulating and preventing heat loss.

The subcutaneous tissue also acts as a cushion, protecting the underlying tissues from physical trauma. In particular, it helps to distribute the forces of impact evenly across the body, protecting the bones and internal organs from damage.

In addition to fat cells, the subcutaneous tissue contains blood vessels, nerves, and lymphatic vessels. These structures provide nutrients and oxygen to the cells of the subcutaneous tissue, as well as help to remove waste products and maintain immune function.

Adipose tissue is composed of fat cells containing stored lipids in the form of droplets. There are multiple small droplets in each cell for lean or normal humans, while for obese humans, there is usually a single large drop. These spherical droplets are evenly distributed within adipocytes. The diameters of the adipocytes range from 15 µm to 250 µm, with a mean diameter of 50 µm to 120 µm. In the spaces between the cells, there are blood vessels, nerves, and reticular fibrils that provide metabolic activity.

A.2.4 Blood and Blood Vessels of the Skin

Human whole blood comprises 55 vol% plasma (90% water, 10% proteins) and 45 vol% cells (99% red blood cells, 1% leukocytes, and thrombocytes). Red Blood Cells (RBC) or erythrocytes are the primary components of blood. The RBC has a characteristic flat biconcave form with a diameter of 7–8 µm and a thickness of 2 µm. RBCs are smaller than other cells, including white blood cells (12µm). This small size allows RBC to fit through the microvasculature easily. Each erythrocyte contains approximately 260 million hemoglobin molecules, which occupy 95% of its volume.[5]

The microvasculature is a network of small blood vessels responsible for providing oxygen and nutrients to the cells in various human body organs and removing waste products. The organization of the microvasculature in the skin is consistent across different parts of the body and age groups, with only slight variations in the density of capillary loops and ascending arterioles. The main components of the microvasculature in the skin are capillary loops, an upper plexus located in the papillary dermis, a lower plexus located at the dermal-subcutaneous interface, and arterioles and venules connecting them. The upper plexus is a horizontal network of capillaries that supplies nutrients to the dermal papillae. In contrast, the lower plexus gives rise to ascending arterioles that connect to the upper plexus. Most of the microvasculature in the skin is located in the papillary dermis, just 1–2 millimeters below the surface of the epidermis. The arterioles in the papillary dermis vary from 17μm to 26μm in diameter and represent terminal arterioles.

The capillaries in the skin, which are the smallest blood vessels, originate from a terminal arteriole in the upper layer of the dermis. They form a loop-like structure of an ascending branch, a horizontal section with a hairpin turn, and a descending branch that connects to a postcapillary venule in the lower layer of the dermis. The diameter of the capillaries is relatively small, with an outer diameter of 10–12 μm and an inner endothelial tube diameter of 4–6 μm.[6]

The branching pattern of the arterioles that supply blood to the capillaries is not uniform and occurs at random intervals of 1.5–7mm.[7] Each arteriole branches into four to five smaller branches that form the upper horizontal plexus of blood vessels. The blood flow pattern in this network can be considered to have an umbrella shape, where the paired ascending arteriole and descending venule represent the handle.

The ultrastructure of the arterioles and venules in the reticular layer differs from the ultrastructure of comparable vessels in the superficial horizontal plexus. In particular, their diameters are wider, 50μm vs. 25μm, and the walls are thicker, 10–16μm vs. 4–5μm.[8] The arterioles and venules in the adipose tissue are identical in structure and size to those of the lower horizontal plexus.

The body has various mechanisms to control the blood flow in small blood vessels of microcirculation. These mechanisms include the autonomic nervous system, metabolic processes, and muscle contractions. However, the primary function of microcirculation regulation is to control body temperature (thermoregulation).

There are some variations in microcirculation between hairless and hairy skin. In hairless skin, there are connections between arterial and venous compartments called Arterio-Venous Anastomoses (AVAs). In hairy skin, a ring of smooth muscle cells surrounds the branching point of the small blood vessels, forming a sphincter-like structure. The network of small blood vessels, called arterioles, plays a key role in determining the amount and distribution of blood flow to the capillaries.

A.3 WOUND MORPHOLOGY

Wound morphology refers to the physical appearance and characteristics of a wound. It is an important aspect of wound assessment and management, as the appearance of a wound can provide valuable information about the stage of the healing process, the presence of infection, and the need for additional treatment.

Several factors can influence a wound's appearance, including the wound's size and depth, the presence of foreign bodies and other tissue types, such as slough or eschar (dead tissue). One of the key factors used to classify wounds is the stage of the healing process. There are four stages of wound healing: hemostasis (bleeding control), inflammation, proliferation (tissue growth), and maturation (tissue repair). The appearance of a wound can vary significantly depending on which stage of the healing process it is in. For example, a wound in the hemostasis stage may appear as a fresh, bleeding wound, while a wound in the maturation stage may appear as a fully healed scar.

Other factors that can affect the appearance of a wound include the presence of infection and the presence of other tissue types. An infected wound may appear red, swollen, and painful and may have a foul smell. A wound that contains slough or eschar may appear yellow or tan and may have a wet or dry appearance. To properly assess and manage a wound, it is important to carefully examine its morphology and consider the various factors influencing its appearance, which can help to identify any potential issues or complications and guide the selection of appropriate treatment options.

A.3.1 Wound Tissue Types

Several types of wound tissue can be present in a wound. These include:

Epithelial Tissue: the epithelium appears light pink and has a shiny, pearl-like appearance. Epithelial cells migrate from the outer edges of the wound and move across the wound bed until the wound is closed.

Granulation Tissue: it is a pink, spongy tissue formed during healing. It comprises new blood vessels, collagen, and other cells. Granulation tissue helps to fill in the wound and provides a framework for the growth of new skin.

Necrotic (nonviable) Tissue: it is the tissue that has died due to a lack of blood flow. Necrotic tissue can be black, brown, or yellow in color and may have a foul smell. Removing necrotic tissue from a wound prevents infection and promotes healing.

Eschar: this is a thick, blackened layer of dead tissue that forms over a wound. It is often the result of a burn or a severe infection.

Slough: it is a yellow or greenish-yellow coagulum often found in wounds with a lot of drainages. It comprises dead white blood cells and is a sign of infection or tissue death. As such, it is not truly a tissue.

Often, nonviable tissue is a term used collectively to describe different types of necrotic tissue. For example, necrotic tissue that undergoes hardening and turns black is known as hard black eschar. As autolysis progresses, the texture and color of the tissue change, resulting in the formation of a soft brown eschar, which eventually transforms into either an adherent grey/yellow slough or a loosely adherent slough. Sometimes, necrotic tissue includes eschar, but slough is considered a different tissue type.

Proper wound care involves regularly inspecting and cleaning the wound to remove any dead or infected tissue and promote the growth of new, healthy tissue, which can help to speed up the healing process and reduce the risk of complications.

A.3.1.1 Eschar

An eschar is a tissue type that forms over a wound or injury. It is a thick, blackened layer of tissue that is often dry and crusty, typically resulting from a thermal or chemical burn. Eschars are also known to form after certain types of infections, such as cutaneous anthrax or botulism.

It is important to note that the terms 'eschar' and 'scab' are not interchangeable. Eschar specifically refers to dead tissue present in a full-thickness wound, whereas the term 'scab' describes a crust that forms by coagulating blood or exudate. Scabs are typically found on superficial or partial-thickness wounds.

The formation of an eschar is a normal part of the healing process, as it serves to protect the wound from further damage and prevents the spread of infection. However, in some cases, eschars may cause problems or complications. For example, an eschar may become too large or extend into surrounding tissue, leading to scarring or disfigurement. Additionally, an eschar may restrict movement or cause discomfort if it forms over joints or other areas of the body that experience a lot of movement.

An escharotomy is a surgical procedure performed in emergencies, which involves making incisions through areas of burnt skin to release the eschar and alleviate its constrictive effects. This procedure can help to restore distal circulation and improve ventilation. It is used to treat full-thickness circumferential burns.

There are several different treatment options for eschars, depending on the location and severity of the wound. Some of the most common treatments for eschars include wound care, skin grafting, and debridement. Wound care involves cleaning the wound and keeping it moist to promote healing. The process of skin grafting involves harvesting healthy skin from a different area of the body and transplanting it onto the area affected by the eschar. This procedure can expedite the healing of the wound. Debridement involves removing dead or damaged tissue from the wound to allow for faster healing. Several types of debridements can achieve the removal of devitalized tissue. These include surgical, biological, enzymatic, and autolytic debridements.

A.3.1.2 Slough

Slough, in the context of wound healing, refers to a 'coagulum' that forms on the surface of a wound as part of the body's natural healing process. Slough is typically yellow or tan and has a wet or moist appearance. It comprises dead cells, bacteria, and other debris accumulated in the wound.

Slough can form in a wound for a number of reasons, including infection, tissue necrosis (death), or the presence of foreign matter in the wound. It can also indicate impaired circulation or an underlying health condition hindering the healing process.

If left untreated, slough can interfere with the healing process and may even lead to further complications, such as infection. Therefore, the slough needs to be removed from the wound to facilitate the healing process. This can be done through wound debridement, which removes dead or damaged tissue from a wound. Debridement can be performed surgically or with specialized wound dressings or topical solutions.

A.3.1.3 Granulation tissue

Granulation tissue is a type of pink, spongy tissue that forms during the healing process of a wound. It is made up of new blood vessels, collagen, and other cells and serves to fill in the wound and provide a framework for the growth of new skin.

The formation of granulation tissue is a sign that a wound is healing properly. It typically appears 2–3 days after a wound has been sustained and is most commonly found in closed or partially closed wounds. Granulation tissue can be encouraged to grow by keeping the wound clean and moist, which can be achieved through the use of moist wound dressings, such as hydrocolloids or hydrogels. These dressings help to maintain a moist environment in the wound, which is essential for the growth and development of granulation tissue.

Granulation tissue is often fragile and prone to bleeding, so it is important to handle it gently and avoid activities that may cause trauma to the wound. In some cases, granulation tissue may need to be trimmed or removed if it becomes too overgrown or blocks the healing process. Proper wound care and management are essential for developing healthy granulation tissue and successful wound healing. They may involve regular cleaning and dressing changes and using medications to control infection and inflammation.

A.3.1.4 Epithelial Tissue

Epithelial tissue forms the outer covering of the body and its organs and the lining of the body's cavities and tubes. It comprises cells that are tightly packed together and arranged in a single layer or multiple layers. In the context of wound healing, epithelial tissue refers to a thin layer of epithelial cells covering the wound and is an integral part of the body's natural repair process.

The regeneration of the epidermis over a partial-thickness wound surface, or scar tissue formation on a full-thickness wound, is referred to as epithelialization. The epithelial cells migrate to the site of the wound and begin to divide, forming a new layer of epithelial tissue (proliferation). As the new epithelial cells continue to divide and mature, they begin to form a layer of tightly packed and sealed cells. The blood clot often provides a scaffold for the migration and proliferation of cells involved in the repair process.

Once the epithelial tissue has fully repaired the wound, the healing process is complete. However, the repair process can be disrupted by multiple factors, such as infection, chronic illness, or the presence of foreign bodies in the wound. In these cases, the healing process may be slowed or impaired, and additional treatment may be necessary to facilitate the repair of the damaged epithelial tissue.

A.3.2 Other Tissues

Some other tissues can be exposed in the case of the wound. They include muscles, nerves, and various types of connective tissues: tendons, ligaments, cartilage, bones, and adipose tissue.

A.3.2.1 Muscles

Muscles are vital in supporting the skeletal structure and facilitating body movement. Healthy striated muscles are characterized by their bright red color due to the abundance of blood vessels present throughout the muscle tissue. Additionally, these muscles have a firm and elastic feel when touched or palpated.

The smallest structural and functional subunits of skeletal muscle are myofibrils that lie parallel to the long axis of the muscle fiber. The diameter of myofibrils is about 1–2 μm, and the resting length of a sarcomere is about 2–3 μm long.[9]

The basic units of the skeletal muscles are muscle fibers. The diameter of a mature skeletal muscle fiber ranges from 10μm to 100 μm. Three layers of connective tissue protect skeletal muscle: the epimysium, the outermost layer, encircles the entire muscle; the perimysium surrounds groups of 10 to 100 muscle fibers, separating them into bundles called fascicles; and the endomysium separates individual muscle fibers from one another.

The optical properties of muscle depend on measurement orientation due to the anisotropic nature of light propagation in muscle. For example, the relative difference in the measured absorption coefficients between the 0° and the 90° probe orientations is as much as 50%.[10]

A.3.2.2 Nerves

Nerves are whitish cordlike structures composed of one or more bundles (fascicles) of myelinated or unmyelinated nerve fibers, coursing outside the central nervous system by which stimuli are transmitted from the central nervous system to a part of the body (motor) or the reverse (sensory). Nerves contain connective tissue within the fascicle and around the neurolemma of individual nerve fibers (endoneurium), around each fascicle (perineurium), and around the entire nerve and its nourishing blood vessels (an epineurium). Nerves comprise the peripheral nervous system, as distinguished from the central nervous system (brain and spinal cord). Nerves are grayish-white and situated within the subcutaneous tissue layer, making them particularly vulnerable to unintended damage.

A.3.2.3 Adipose tissue

Adipose tissue, commonly known as fat, is primarily located in the subcutaneous layer of the skin and typically appears as a globular, yellowish-white mass (except for newborns, when it may be brown in color). However, if the adipose tissue becomes damaged, its color may darken to a deeper shade of yellow.

A.3.2.4 Bones

Bone is a rigid and strong connective tissue that makes up the body's skeleton. It comprises cells called osteocytes, surrounded by a hard matrix of minerals. When exposed, bones appear white or pale yellow in color and have a hard texture when healthy. Bones are often overlooked during visual assessments, so it may be necessary to gently probe the wound base with a wooden end of a cotton-tipped applicator or palpate the area with a gloved finger to confirm the presence of hard bone.

A.3.2.5 Tendons and Ligaments

Dense connective tissue is found in areas of the body where there is a need for strength and stability, such as tendons and ligaments. It comprises closely packed collagen or elastic tissue fibers and is typically less flexible than loose connective tissue. Tendons and ligaments connect two other tissues together (i.e. bone to muscle or bone to bone, respectively). Their color can range from white to yellow, depending on the hydration level. Healthy tendons are shiny and white. However, when damaged or dehydrated, it darkens to a yellowish color as it begins to die. Therefore, maintaining moisture in tendons is crucial.

A.3.2.6 Callus

Corns and calluses are thickened skin areas that develop due to repeated friction, pressure, or other types of irritation. They are formed by accumulating terminally undifferentiated keratinocytes in the outermost layer of skin. Although the cells of calluses are dead, they are highly resistant to mechanical and chemical damage due to extensive networks of cross-linked proteins and hydrophobic keratin intermediate filaments that contain many disulfide bonds. Calluses are a natural response to skin irritation on the palms or soles of the feet. However, excessive friction that occurs too quickly for the skin to develop a protective callus can cause blisters or abrasions instead.

The formation of a callus is a complex process that involves several biochemical and mechanical changes in the skin. First, the mechanical stress activates keratinocytes in the epidermis, which begin to divide and differentiate faster, which leads to the formation of a thicker layer of keratin, the protein that makes up the stratum corneum. Additionally, mechanical stress also causes the accumulation of extracellular matrix molecules such as collagen and glycosaminoglycans, which contribute to the thickening of the stratum corneum.

Several risk factors, such as foot deformities (e.g. bunions, hammertoe) and not wearing socks or protective gloves, can contribute to the formation of calluses. While calluses are usually harmless, if left untreated, they can lead to skin ulceration or infection, which is particularly important for patients with diabetes. They can also cause patients to shift their weight off the affected, painful area, placing excessive stress on the asymptomatic side.

Therefore, the removal of calluses is an essential part of surgical debridement. However, healthcare professionals often have difficulty identifying calluses, which are often unsightly and not as clearly visible as corns. So some areas of dead skin can be missed during the debridement process.

NOTES

1. GF Odland, "Structure of the skin", in *Physiology, Biochemistry, and Molecular Biology of the Skin, Vol. 1*, L. A. Goldsmith (ed.), pp. 362, Oxford University Press, Oxford (1991).
2. SL Jacques, "Origins of tissue optical properties in the UVA, visible and NIR regions", in RR Alfano, JG Fujimoto (editors), *Advances in Optical Imaging and Photon Migration*, OSA TOPS: Optical Society of America, Vol. 2, Washington, DC, 1996, pp. 364–371.
3. AN Bashkatov, EA Genina, VV Tuchin, "Optical properties of skin, subcutaneous, and muscle tissues: A review", *J Innov Opt Health Sci*, 4(1): 9–38 (2011).
4. IV Meglinski, SJ Matcher, "Quantitative assessment of skin layers absorption and skin reflectance spectra simulation in visible and near-infrared spectral region", *Physiol Meas*, 23: 741–753 (2002).
5. T Yoshida, M Prudent, A D'Alessandro, "Red blood cell storage lesion: Causes and potential clinical consequences", *Blood Transfus*, 17(1):27–52(2019).
6. IM Braverman, "The cutaneous microcirculation", *J Invest Dermatol*, 5(1): 3–9(2000).
7. IM Braverman, J Schechner, "Contour mapping of the cutaneous microvasculature by computerized laser Doppler velocimetry", *J Invest Derrnatol*, 97(6): 1013–1028 (1991).

8. IM Braverman, A Keh-Yen, "Ultrastructure of the human dermal microcirculation. III. The vessels in the mid- and lower dermis and subcutaneous fat", *J Invest Dermatol*, 77: 297–304 (1981).
9. HJ Swatland, "Relationship between pork muscle fiber diameter and optical transmittance measured by scanning microphotometry", *Can J Anim Sci*, 82: 321–325 (2002).
10. G Marquez, LV Wang, S Lin, et al. "Anisotropy in the absorption and scattering spectra of chicken breast tissue", *Appl Opt*, 37: 798–804 (1998).

APPENDIX B

Bio Optics Primer

While interacting with biological tissues, light experiences a broad range of phenomena, which can be used for therapeutic and diagnostic purposes.

The primary phenomena are absorption and refraction, which give rise to a wide range of other phenomena, including fluorescence and scattering.

In this Appendix, we briefly overview the basic physics of light and light interaction with skin that defines light propagation in the skin. Light refraction, scattering, absorption, and spectral and polarization properties are analyzed.

B.1 PROPERTIES OF LIGHT

Electromagnetic (EM) waves are transverse waves. As such, the electric and magnetic fields, which are perpendicular to each other, change (oscillate) in a direction perpendicular to the wave's propagation direction.

Numerous factors impact light interactions with tissues. Some of them, including photon energy (frequency or wavelength), coherence, bandwidth, and polarization, can be considered fundamental properties of light. We will briefly discuss them. However, multiple other factors impact interaction with tissue, which are important, particularly in therapeutic applications:

- Power or energy of the incident light;
- Spot size (irradiated area);

- Irradiance or radiant exposure (power or energy per unit area, respectively);
- Duration of the irradiation (the pulse duration);
- Spatial profile (how irradiance varies across the beam);
- Temporal profile (how irradiance varies with time during the pulse).

B.1.1 Spectral Ranges

The electromagnetic spectrum refers to the immense range of electromagnetic waves from 1 Hz to 10^{25} Hz. It is split into several (typically seven) regions or bands; radio waves, microwave radiation, infrared radiation, visible light, Ultraviolet radiation, X-rays, and gamma radiation. However, this separation is quite nominal, as there are no precisely defined boundaries between the electromagnetic spectrum bands; rather, they morph in each other, and the radiation of each band has a mix of properties of the two regions of the spectrum that bound it. Moreover, boundaries between regions and sub-regions are tentative, as different authors and agencies may use slightly different definitions.

The optical range is a subset of the overall electromagnetic spectrum. According to DIN 5031, the term 'optical radiation' refers to electromagnetic radiation in the wavelength range between 100 nm and 1 mm, encompassing Ultraviolet radiation, visible light, and infrared radiation. However, occasionally this range is extended to include the 10–100nm range, which belongs to Vacuum UV (V-UV) radiation. Sometimes, only a range perceived by the human eye (380–760nm) is called the optical range.

Each region in the optical range can be split into several sub-regions.

The infrared part of the electromagnetic spectrum covers the range from roughly 1 mm to 750 nm. The primary interaction mechanism with biological tissues in this range is the excitation of molecular rotations and vibrations. The infrared range is typically divided into three parts:

Far-infrared: 1 mm – 10 µm. Far-infrared radiation is strongly absorbed by gases (rotational modes of molecules), liquids (molecular motions), and solids (phonons).In particular, the water strongly absorbs in this range, which impedes propagation through the atmosphere. The low energy part of this range is often called microwaves or terahertz waves.

Mid-infrared: 10–2.5 µm. The mid-infrared spectrum range, which is relevant to biological tissues, is characterized by significant thermal emission, with human skin at normal body temperature emitting strongly at the lower end of this region. Additionally, this spectrum range is also characterized by strong absorption, as mid-infrared radiation excites molecular vibrations. Therefore, it is often referred to as the fingerprint region, as a compound's mid-infrared absorption spectrum is unique to that particular compound.

Near-Infrared or NIR: 2,500–750 nm. Physical processes relevant to this range are governed by molecular vibrations (as in other IR ranges) and electron excitations (as visible light).

Visible light. The human eye detects electromagnetic waves with wavelengths between 380 nm and 760 nm. Its 400–750nm sub-range, referred to as visible light, is perceived as different colors, including violet, 400–450 nm; blue, 450–480 nm; green, 510–560 nm; yellow, 560–590 nm; orange, 590–620 nm; and red, 620–750 nm. Molecular electron excitation is the primary interaction mechanism with biological tissues in this range.

Ultraviolet radiation covers the wavelength range from 10nm and 400nm. Molecular and atomic valence electron excitation is the primary interaction mechanism with biological tissues in this range. The Ultraviolet region is typically split into four ranges.

Ultraviolet A or UVA: 315–400nm. Sun radiation in this range is mostly not blocked by the atmosphere and accounts for 95% of UV radiation at the Earth's surface. Moreover, it can penetrate the deeper layers of the skin and is responsible for the immediate tanning effect.

Ultraviolet B or UVB: 280–315nm. The ozone layer in the atmosphere mostly blocks the radiation from the sun in this range. UVB radiation is highly biologically active, but it is unable to penetrate beyond the topmost layers of the skin. It is responsible for both delayed tanning and burning, and the photons it emits can directly damage DNA. Specifically, it excites DNA molecules in skin cells, forming abnormal covalent bonds between adjacent pyrimidine bases, resulting in a dimer.

Ultraviolet C or UVC: 200–280nm. However, some sources (e.g., ISO 21348 or WHO) extend it to 100nm. The ozone layer in the atmosphere completely blocks the sun's radiation in this range. This range causes photochemical reactions and is used extensively for germicidal purposes.

Vacuum Ultraviolet, or V-UV: 10–200nm. The name for this range comes from the fact that radiation is entirely absorbed by gases in the air. Thus, it can be observed only in a vacuum. The radiation with the longest wavelength capable of ionizing atoms and causing chemical reactions by separating electrons from them is known as Vacuum UV. Ultraviolet radiation with a wavelength shorter than 121.6nm (hydrogen Lyman-alpha line) is called extreme Ultraviolet (E-UV) and is considered ionizing. For example, US FCC defines 10eV (124nm) as the threshold for ionizing radiation.

At energy lower than 10eV, photons do not ionize atoms but may break bonds, making molecules more reactive. It is the case for the Far UV range of V-UV (121.6–200nm), UVC (200–280nm), and UVB (280–315nm). However, even UVA may create oxygen radicals, mutations, and skin damage. Not surprisingly, during the 19th century, UV radiation was called 'chemical rays.'

This chapter will focus on a subset of the optical range, encompassing the visible light and Near-Infrared (NIR) ranges. In addition, we briefly consider mid-wave and long-wave IR, which are used in thermography.

B.1.2 Coherence

Electromagnetic waves can retain a consistent phase relationship over a certain period of time, known as the coherence time. This phase relationship remains the same for any point in the plane perpendicular to the propagation direction.

Coherence is a property typically associated with laser light sources. The degree of temporal coherence of the emitted light defines the coherence length of a light source, $l_C = c\tau_C$, where c is the light speed, and τ_C is the coherence time. The coherence time is approximately equal to the pulse duration of the pulsed light source or inversely proportional to the wavelength bandwidth $\Delta\lambda$ of a continuous wave light source, $\tau_C \sim 0.44 \lambda^2/(c \Delta\lambda)$.

The coherence length is a crucial factor in Optical Coherence Tomography (OCT), as it determines the achievable image resolution. A shorter coherence length, l_C, corresponds to a higher resolution. For example, OCT systems that use a titanium sapphire laser or a white light source can provide a subcellular resolution of 1–2 µm. In particular, the titanium sapphire laser with λ = 820 nm $\Delta\lambda$ can be up to 140 nm; therefore, the coherence length is very short, $l_C \approx 2$ µm.

B.1.3 Monochromatic vs. Polychromatic

Monochromaticity refers to the wavelength bandwidth of the emitted light. In particular, monochromatic light is a single color (wavelength) light, which to some extent is an idealization, as there are no light sources with a zero optical bandwidth (although continuous-wave single-frequency well-stabilized lasers can have bandwidth below 1Hz).

A more realistic scenario is quasi-monochromatic light, which refers to light with narrow but nonzero wavelength bandwidth. Quasi-monochromatic light can be presented as a group of monochromatic waves with slightly different wavelengths. In addition to lasers, certain gas discharge lamps and metal vapor lamps (e.g., mercury vapor lamps and sodium vapor lamps) emit light in narrow spectral lines.

Non-monochromatic light is light with a broad wavelength bandwidth, which can be presented as many monochromatic waves with different wavelengths.

B.1.4 Polarization

Light, like any electromagnetic wave, is a traverse wave. Thus, if we consider isotropic optically inactive media (more complex cases will be briefly mentioned later), the electric field vector E oscillates in the direction perpendicular to the wave vector k. This wave's plane containing vectors E and k is called a polarization plane. Any orientation of the electric field vector E can be considered as a superposition of two mutually perpendicular directions. While these directions can be arbitrary during propagation in free space like air or vacuum, interaction with biological tissues typically breaks this symmetry. If we consider non-normal (oblique) incidence light on the tissue, then the natural choice would be to define a plane containing the wave vector k and the normal vector to the surface, n (plane of incidence). In this case, we can consider any wave as a superposition of the wave polarized in the plane of incidence, p-plane (parallel), and the plane perpendicular to the p-plane and wave vector k, s-plane (from *Senkrecht*, German for perpendicular). As a result, when the electric field of polarized light is aligned with the plane of incidence, it is labeled as p-polarized. On the other hand, when the electric field is perpendicular to the plane of incidence, it is referred to as s-polarized. P polarization is sometimes referred to as Transverse-Magnetic (TM) and can also be called pi-polarized or tangential plane polarized. Similarly, s polarization can also be called Transverse-Electric (TE), as well as sigma-polarized or sagittal plane polarized.

The solution for refraction of p- and s-waves on the interface between two media is known as Fresnel equations.

Another way to select a direction is to use a coordinate system associated with a polarization of a light source. For example, for a linearly polarized laser beam or light after a polarizer, we can define light polarization with respect to the initial beam polarization (e.g., as parallel and perpendicular with respect to the initial polarization plane).

When polarized light transverses a tissue, it gradually depolarizes due to the multiple scattering/refraction events in the inhomogeneous (scattering) medium (i.e. tissue). Namely, the destruction of light polarization occurs. This process can be characterized by the degree of polarization (or depolarization ratio), which is the quantity that describes the ratio of the intensity of polarized light to the total intensity of light, $P_L = (I_{\parallel} - I_{\perp})/(I_{\parallel} + I_{\perp})$, where $I_{\parallel}$ is the intensity of light polarized in parallel and $I_{\perp}$ perpendicular to polarization plane.

So far, we have discussed linear polarization. However, suppose two electric field components in orthogonal directions have phase differences between them. In that case, it leads to the rotation of the electric field vector around the direction of the wave (wave vector k). This case is called elliptical polarization. Circular polarization is a particular case of elliptical polarization where the strength of electrical fields across both directions is the same.

By transmitting circularly polarized light through a quarter-waveplate, it can be transformed into linearly polarized light. For instance, if linearly polarized light is transmitted through a quarter-waveplate whose axes are at a 45° angle to its polarization axis, it will change to circular polarization.

Most light sources in nature are non-polarized as emissions from atoms or molecules must be synchronized (coherence) to produce polarized light, as in laser light sources. Non-polarized light can be considered to be a mixture of linearly polarized waves with multiple polarization planes. As such, non-polarized light can be polarized using various optical elements, e.g., polarizers. Polarizers allow passing the light just with one direction of linear polarization.

Biological tissues are known to demonstrate a certain degree of structural anisotropy. For example, they may contain elongated collagen fibers, which can be organized in oriented structures. As a result, many tissues, including skin, demonstrate polarization anisotropy, an inequality of polarization properties along different axes resulting in structural birefringence. Birefringence refers to a material property that causes an incident ray of light to be split into two rays, known as an ordinary ray and an extraordinary ray, which are either linearly polarized in perpendicular planes or circularly polarized in opposite directions. Skin linear birefringence is mainly due to the linear anisotropy of the dermis fibers, as the refractive index is higher along the length of the fibers than across them. The polarization of light passing through the tissue can be affected by changes in the tissue's structure, such as due to collagen aging in the skin.

Some media can be optically active, like chiral media, which rotate the polarization plane of linearly polarized light (e.g., glucose).

B.2 ABSORPTION

Light absorption is the transformation of light (radiant) energy to some other form of energy – usually heat – as the light travels through media. Generally, an absorbing medium contains absorption centers: particles or molecules that absorb light.

The propensity of the medium to absorb light can be characterized by the absorption coefficient, which can be introduced in the following way. It is known that the intensity of the collimated light in the non-scattering media (e.g., liquid) decreases exponentially with the distance. Therefore, if we define the distance L_a as the distance where the intensity of the incident light decreases by e times ($I(L_a) = I_0/e$), the absorption coefficient μ_a can be defined as an inverse value of the distance L_a ($\mu_a = 1/L_a$). Using this definition of the absorption coefficient, we can write the Beer law for any distance l:

$$I(l) = I_0 e^{-\mu_a l} \tag{B.1}$$

B.2.1 Tissue Chromophores

Absorption of the UV and visible light in the skin is due to the electronic excitation of aromatic or conjugated unsaturated chromophores.

Proteins found in the epidermis contain the aromatic amino acids tryptophan and tyrosine, which have a characteristic absorption band near 270–280 nm; urocanic acid and the nucleic acids also contribute to this absorption band with a maximum near 260–270 nm.

Epidermal melanin plays an important role in limiting the penetration depth of light in the skin: it effectively absorbs at all wavelengths from 300 to 1,200 nm, but the strongest absorption occurs at shorter wavelengths, in the near-UV spectral range.

Melanin and hemoglobin are the primary chromophores in the visible range, while water and lipids are the main absorbers in the NIR spectral range.

B.2.1.1 Hemoglobin

The two primary blood chromophores are deoxygenated and oxygenated hemoglobin (Hb and HbO_2). These chromophores are of great clinical interest as they define the total hemoglobin concentration (tHb=Hb+HbO_2) and blood oxygenation (SO_2=HbO_2/(HbO_2+Hb)). Deoxyhemoglobin has absorption peaks at 272, 433, 556, and 758nm.[1]

Oxyhemoglobin has absorption peaks at 274, 344, 414 (Soret band), 542, and 576nm (Q bands) and a broad peak around 940nm.[1]

Hematocrit is the volume fraction of cells within the whole blood volume. A typical range is from 40% to 45%. The hemoglobin concentration ranges from 134 to 173 g/l for whole blood and 299 to 357 g/l for RBCs.[2]

In addition to Hb and HbO_2, other hemoglobin (methemoglobin (metHb), sulfhemoglobin (SHb), and carboxyhemoglobin (COHb)) may also be present in the blood; however, they are in much smaller quantities.

Methemoglobin is a biomolecule that is formed when the iron in hemoglobin is oxidized to its ferric state. In healthy individuals, only 1% of hemoglobin is methemoglobin; anything above this is called methemoglobinemia. The hemoglobin–methemoglobin couple acts as an oxygen tension sensor by reacting with Nitric Oxide (NO). NO dilates vessels and increases perfusion to areas with poor oxygenation. Studies have shown that inadequate wound healing in diabetics is linked to reduced NO production in the wound.[3]

Most of the daily reduction of methemoglobin in healthy people is due to the NADH cytochrome b5/cytochrome b5 reductase system, an efficient antioxidant found in red blood cells.[4] On average, methemoglobin is reconverted at a rate of 15% per hour.

B.2.1.2 Melanin

Melanin is a pigment produced by melanocytes, which can be found in the epidermis. It is responsible for the color of the skin, hair, and eyes, and it also protects the skin from the harmful effects of the sun's Ultraviolet (UV) rays.

There are two main types of melanin: eumelanin and pheomelanin. Eumelanin produces darker colors, such as brown, black, and tan, while pheomelanin produces lighter colors, like red and yellow. The type and amount of melanin produced by an individual are determined by their genetics.

Melanin is produced in response to UV radiation, so people's skin tends to get darker when exposed to the sun. The more melanin is made, the darker the skin appears. This mechanism is a natural way for the body to protect itself from the harmful effects of the sun, such as sunburn, skin cancer, and premature aging.

However, melanin is not equally distributed throughout the body. Some areas of the skin, such as the face, arms, and legs, have more melanin than others, such as the palms of the hands and the soles of the feet.

Even though the number of melanin-producing cells (melanocytes) is similar in all skin types, melanocytes in dark skin are more active than those in light skin, resulting in increased melanin production.

Melanin demonstrates broadband absorption from UV to visible light to NIR. The absorption is strongest in the UV range and gradually decreases with the wavelength increase. However, melanin also is a strong scatterer, which is caused by its high index of refraction (1.6–1.7 in visible and NIR ranges).

B.2.1.3 Water

Water is the most abundant component of biological tissues. It accounts for 60–80% of overall body weight.

In the visible range, water absorption is minimal and can be ignored for practical purposes. However, the water spectrum in NIR has multiple features which can be used in diagnostics and therapy. In particular, water has overtone bands of the O-H bonds with peak absorption at 760 nm, 970 nm (the second overtone of the O-H stretching band), 1,190nm (the combination of the first overtone of the O-H stretching and the O-H bending band), 1,450 nm (the first overtone of the OH-stretching band and a combination band), and 1,940 nm (combination of the O-H stretching band and the O-H bending band).[5]

The absorption coefficient of water at 970, 1,190, 1,450, and 1,940nm wavelengths are 0.45, 1.04, 28.6, and 119.83cm^{-1}, respectively.[6]

B.2.1.4 Fat

Lipids are the main components of cell membranes and, therefore, can be found anywhere in the body. However, they can be found in large quantities in adipose tissues, such as the subcutaneous fat layers and visceral fat surrounding the organs. The lipid concentration in adipose tissue ranges from 60% to 87% for adults and 23% to 47% for infants.[7]

The absorption of animal fat was measured in the 400–2,200nm range.[8] The authors reported multiple peaks in the NIR range, including 930 and 1,040nm (13.1 and 7.0 m^{-1}, respectively). Thus, lipid absorption in realistic biological tissues is generally relatively minimal. The lipid absorption spectrum depends on the proportion of various types of lipids (saturated, monounsaturated, and polyunsaturated fat) in the tissue.

The absorption of major tissue chromophores is depicted in Figure B.1.

B.2.1.5 Other Chromophores

In addition to primary chromophores, multiple other chromophores are present in the skin, including cytochrome c-oxidaze, β-carotene, and bilirubin.

Cytochrome c. Cytochrome c oxidase (or Complex IV) is a respiratory energy-transducing enzyme, the terminal electron acceptor of the mitochondrial respiratory chain. It is responsible for over 90% of cellular oxygen consumption and essential for the efficient generation of cellular adenosine triphosphate (ATP). While the total concentration of cytochrome c does not change in time, its redox state changes, which can be used to monitor tissue hypoxia. The enzyme contains four redox active metal centers; one of these, the binuclear Cu_A center, has a strong absorbance in

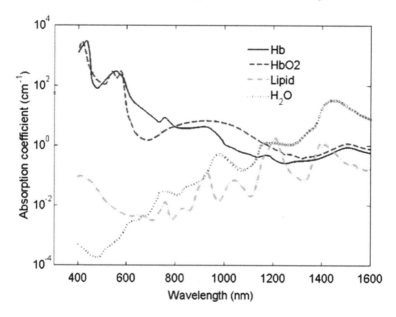

FIGURE B.1 The absorption of major tissue chromophores in visible and NIR ranges.

the Near-Infrared, enabling it to be detectable in vivo by Near-Infrared spectroscopy. In particular, the cytochrome oxidase extinction coefficient is high in the oxidized form (>3 mM^{-1} cm^{-1}).[9] However, only oxidized Cu$_A$ demonstrates strong absorption at 830nm, while the redox state demonstrates no significant absorption. For example, this difference can be used to assess the redox state in neonatal brain monitoring. However, due to its low concentration with the current state of technology, it is most likely impractical to use cytochrome c to monitor redox in the skin. However, cytochrome c may play an important role in therapeutic applications, including wound healing. For example, photobiomodulation or Low-Level Laser (or Light) Therapy (LLLT) by light in the red to Near-Infrared range (630–1,000 nm) using Low Energy Lasers or LEDs has been shown to accelerate wound healing, improve recovery from ischemic injury in the heart, and attenuate degeneration in the injured optic nerve. It can be attributed, at least in part, to intracellular signaling mechanisms triggered by the interaction of NIR light with the mitochondrial photoreceptor molecule cytochrome c oxidase.[10]

β-carotene. β-carotene is a carotenoid found in the skin, bloodstream, liver, and adipose cells. It is thought that more than 80% of β-carotene reserves are found in adipocytes.[11] β-carotene, along with other carotenoids (α-, γ-, β-carotene, lutein, zeaxanthin, lycopene, and their isomers), can be found in the skin (0:22–0:63 nmol/g)[12] where it protects the living cells from oxidation. This chromophore has its main absorption peak at 482 nm. Its peak position and width depend on the substance in which it was diluted. However, it has low concentrations in the skin (typically on the scale of 0.5 nmol/g),[13] which makes it impractical to use any spectroscopy other than Raman.

Bilirubin. Bilirubin is a red–orange compound produced in the normal catabolic pathway of heme breaking down. It is mainly found in the spleen and liver. Excess bilirubin secretion in the liver can no longer be stored in bile and therefore flows into the bloodstream and leaks into the tissue, causing the skin to appear yellow. This chromophore has its main absorption peak at 467 nm. However, its peak's position and width depend on the substance it was diluted in.

B.2.2 Energy Dissipation Mechanisms

After a photon has given its energy to the absorption center (e.g., molecule), there are multiple ways for it to return to its lowest energy state (ground state), the most common state at room temperature. Light absorption by tissues may induce radiation emission such as fluorescence or initiate energy confinement leading to temperature rise. Finally, the absorbed energy may drive a chemical reaction (intra- and intermolecular energy transfer, isomerization, dissociation, and ionization).

If it results in photon emission, the process is known as luminescence. Fluorescence and phosphorescence processes are two primary types of luminescence. They are illustrated in the Jablonski diagram (see Figure B.2).

When a photon is emitted from an excited state to a lower energy state with the same spin state, for example, $S_1 \rightarrow S_0$, the process is referred to as fluorescence. On the other hand, if the emission occurs between different spin states, for example, $T_1 \rightarrow S_0$, it is called phosphorescence. Fluorescence is much more likely to occur than phosphorescence, resulting in shorter lifetimes for fluorescent states (ranging from 10^{-6} to 10^{-10} seconds) compared to longer lifetimes for phosphorescent states (ranging from 10^{-4} seconds to minutes or even hours). Additionally, nonradiative processes can occur, including internal conversion, intersystem crossing, and vibrational relaxation. These processes can be seen in the diagram, where internal conversion is the transition between energy states of the

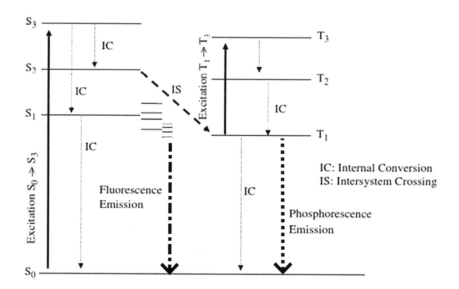

FIGURE B.2 Jablonski diagram. Absorbed energy from S_3 can return to the ground state via Internal Conversions (IC, with or without light emission) or Intersystem Crossing (IS) to the triplet state.

same spin without the emission of radiation. Intersystem crossing is the transition between different spin states without the emission of radiation.

B.2.2.1 Fluorescence

Fluorescence is a property of emitting light of a longer wavelength on the absorption of light energy, which essentially co-occurs with the excitation of a sample. Unlike inelastic scattering, the fluorescence output strongly depends on the energy (wavelength) of the incident light. As such, fluorescence is characterized by both the emission and excitation spectra. In particular, an emission spectrum represents the light emitted by a luminescent material at various wavelengths when a specific, limited range of shorter wavelengths stimulates it. The excitation spectrum is the emission spectrum monitored at one wavelength, with the intensity at that wavelength being measured in relation to the wavelength used to excite it.

Fluorescence effectiveness is characterized by quantum yield, QY, defined as the proportion of photons emitted by fluorescence compared to the photons absorbed.

Fluorophores can be broadly classified into three major types: inorganic molecules (called pigments), organic molecules (called dyes), and nanostructures/nanoparticles.

Autofluorescence (AF) is a natural fluorescence of a material (a tissue) due to the excitation of the endogenous fluorophores in contrast to the fluorescence of a stained material (a tissue or a cell) when exogenous fluorophores are excited.

Native fluorophores in human skin include collagen, elastin, amino acids (tryptophan and tyrosine), NADH, NAD, FAD, and porphyrins. Epidermis contains more tyrosine and tryptophan than the whole skin, so it has a high Autofluorescence in the UVA range. Similarly, the psoriatic stratum corneum has a much higher fluorescence than the normal one.

In addition to native fluorescence, some bacteria species produce fluorophores, indicating bacteria presence. In particular, *P.aeruginosa* produces siderophore pyoverdin, which fluoresces in the blue–green range (peak at 465nm), while more than 20 other species (including clinically relevant *S. aureus*) produce porphyrins, which fluoresce in orange–red range (600–660nm).

Several external fluorescent agents, like fluorescein or Indocyanine Green (ICG), are used to visualize perfusion in tissues using exogenous fluorescence. In the fluorescent angiography procedure, the dye is injected into the bloodstream. Then the target area is illuminated with the excitation light, and the imaging sensor detects the emitted light. Initially, the method was developed for ophthalmology. Fluorescent Angiography (FAG) or Intravenous Fluorescein Angiography (IVFA) is a method that involves a specialized camera and a fluorescent dye to investigate the circulation of the retina and choroid. Recently, the technique has been extended to other blood vessels and body parts.

B.2.2.2 Photoacoustics

If the absorbed energy is not reemitted in the process of luminescence, it will be transferred to other systems, which may raise the local temperature. In turn, the temperature rise can lead to multiple phenomena, from sound wave generation to ultimate tissue destruction. For example, Photoacoustic Tomography (PAT) uses pulsed laser light to irradiate tissues. As a result, pressure waves are produced due to the increased temperature and volume.[14] The process of photoacoustic signal generation can be described in three steps: (1) an object absorbs light, (2) the absorbed optical energy is converted into heat and generates a temperature rise, and (3) thermoelastic expansion takes place, resulting in the emission of acoustic waves.[15] For effective PAT signal generation, the laser pulse duration should be less than both the thermal and stress confinement times.[16] Thus, the typical laser pulse duration used in PAT is on scales of several nanoseconds or less.

A high-frequency ultrasound transducer monitors these pressure waves, and a 3-D reconstruction is performed. Visible and Near-Infrared light is used to perform PAT. The contrast in PAT images depends mainly upon the absorption properties of tissues.[17] The optical absorption is specific to tissue chromophores, and by changing the wavelength of light, one can tune the images to obtain enhanced contrast for particular chromophores.

By projecting light at hemoglobin's peak absorption, inflamed, hyperemic tissue appears dark (hypoechoic) on PAT, while surrounding tissues reflecting such light waves appear bright (hyperechoic).

B.2.2.3 Chemical Reactions

Depending on the photon energy, it may cause a wide range of photochemical reactions. For example, the excited molecule can transfer energy to another part of the molecule (intramolecular energy transfer). This process is called intermolecular energy transfer if the energy is transferred to another molecule. This type of reaction is integral to Photodynamic Therapy (PDT), where the photon is absorbed by one molecule (photosensitizer), and the energy is transferred to another molecule (Reactive Oxygen Species or ROS) that can induce desired action.

In addition, there are light-induced isomerizations (for example, isomerization of bilirubin during PDT), dissociations, and ionizations (at higher energies).

B.3 REFRACTION

Refraction refers to the change in the direction of the light or radio waves passing through the interface between two media or through a medium of varying density. In this context, refraction is a general term that includes reflection as a special case.

Light refraction can be characterized by a refractive index.

B.3.1 Refractive Index

The absolute index of refraction, n, is the ratio of the speed of light in a vacuum, c, to the speed of light in a medium, v. An interface between two media can be characterized by the relative refractive index. The relative refractive index of medium 2 with respect to another medium 1 (n_{21}) is given by the ratio of the speed of light in medium 1 (v_1) to that in medium 2 (v_2). This can be expressed as $n_{21}=v_1/v_2=n_2/n_1$.

The refractive index may vary with wavelength. For example, this causes white light to split into constituent colors when refracted (dispersion).

B.3.2 Light Propagation Through the Interface between Two Media

Propagation of light through the interface between two media with different absolute indexes of refraction (commonly called mismatched boundary conditions) causes a range of phenomena, including ray bending and reflection.

B.3.2.1 Ray Bending

If we characterize the direction of light by the incidence angle θ (the angle with respect to the vector normal to the surface), then light propagation can be described by Snell's law: $n_1 sin\theta_1 = n_2 sin\theta_2$. As a result, when light moves into a material with a higher refractive index, it will bend toward the normal to surface. The degree of deviation toward the normal direction will be greater with a higher refractive index. Conversely, when the light enters a medium with a lower refractive index, it will curve away from the normal and toward the surface.

B.3.2.2 Ray Reflection

In addition to bending, light rays on the interface between two media experience the reflection. Fresnel equations give the general solution to light propagation on the interface between two media. Again, we can consider two cases here. When light enters a material with a higher refractive index, the light will be reflected partially (specular reflection). In the case of normal incidence, the specular reflectivity can be found as $r_s = \left|\dfrac{n_1 - n_2}{n_1 + n_2}\right|^2$.

In general, the light reflection depends on the incident wave polarization.

For realistic tissues ($n=1.33-1.45$), typical values of specular reflection r_s are pretty low, in the 2–4% percent range. The specular reflection is typically an undesirable phenomenon that must be avoided or mitigated (e.g., oil immersion). For example, in most imaging applications, even such a low value of specular reflection coefficient can pose a significant problem of hot spots and mask the useful signal. However, the specular reflection can be used to detect changes in skin surface geometry.[18]

When passing into a medium with a lower refractive index (n_1), the light behavior will depend on the angle of incidence with respect to the critical angle ($\theta_c = arcsin(1/n_{21})$). For small angles of incidence ($\theta < \theta_c$) the light will experience partial reflection (specular reflection). For the angle of incidence bigger than the critical angle ($\theta > \theta_c$), the light will experience the total internal reflection.

The described phenomena have interesting implications in biooptics. In particular, it is known that the light intensity just below the surface of biotissues can be higher (on a scale of 2–4) than the intensity of the incident light.[19]

Proteins, water, and lipids are the primary contributors to the index of refraction of biotissues. While lipids are located primarily in cell membranes and adipose tissues, water and proteins determine the bulk index of refraction of skin layers other than subcutaneous fat. The index of refraction measurements in the human skin was summarized in.[20] For example, the 850–1,064nm measurements yield values in the range of 1.417 to 1.43 for the epidermis.[21] The dermis contains more water; thus, its index of refraction is slightly (typically by 0.01–0.05)[21] less, which will result in almost negligible scattering on the dermis/epidermis interface (<<1%).

A mismatch of refractive indexes between cell membranes (lipids) and cytoplasm (water and proteins) causes light scattering on cells, which is one of the primary light-scattering mechanisms in biotissues.

B.4 SCATTERING

Light scattering refers to a broad range of physical processes where electromagnetic waves are forced to deviate from a straight trajectory by localized non-uniformities (including particles and radiation). Light scattering in biological tissues occurs predominantly on the intracellular and extracellular objects with mismatched indexes of refraction: nuclei, organelles, collagen fibers, etc.

Similar to absorption, it can be described phenomenologically by introducing the coefficient of scattering of the medium. Let's consider the propagation of the collimated light in a non-absorbing medium and define the distance L_s as the distance where the intensity of the incident light decreases by e times ($I(L_s) = I_0/e$). The coefficient of scattering μ_s will be an inverse of the distance L_s ($\mu_s = 1/L_s$). Using this definition of the scattering coefficient, we can write the expression similar to the Beer law:

$$I(l) = I_0 e^{-\mu_s l} \tag{B2}$$

The scattering by an individual molecule or particle is characterized by a so-called 'scattering cross-section' σ_s. The scattering coefficient of the media μ_s can be linked with the scattering cross section σ_s through the concentration of the scattering centers N: $\mu_s = N\sigma_s$

Based on the energy transfer (or lack thereof) between the photon and the media, the scattering phenomena can be divided into inelastic and elastic scattering, respectively.

B.4.1 Elastic Scattering

The predominant mechanism of electromagnetic wave scattering in turbid media is the so-called 'elastic' scattering. In this type of scattering, the energy transfer between the photon and the scattering center does not occur. As a result, the incident and scattered photons have the same energy. The lifetime, τ_e, of the elastic scattering process, is usually very short on the order of 10^{-15} s.

When a uniform sphere disperses an electromagnetic plane wave, this phenomenon is referred to as the Mie solution to Maxwell's equations. It is also recognized as Lorenz–Mie solution, Lorenz–Mie–Debye solution, or Mie scattering. Its solution is an infinite series of spherical multipole partial waves. It gives an accurate approximation for the scattering of particles, which are much larger or smaller than the wavelength of the scattered light. These regimes are typically referred to as Rayleigh scattering and Mie scattering, respectively.

B.4.1.1 Rayleigh Scattering

Rayleigh scattering is the elastic scattering of light or other electromagnetic radiation by particles much smaller than the wavelength of the radiation ($d<<\lambda$, where d is the size of the particle, λ is the wavelength). In this case, in the absence of the resonant frequency, the cross-section of scattering is inversely proportional to the fourth power of the wavelength (namely, $\sigma_s \sim d^6 / \lambda^4$). The significant dependency of the scattering on the wavelength ($\sim\lambda^{-4}$) results in the scattering of shorter (blue) wavelengths being stronger compared to longer (red) wavelengths, which explains the blue color of the sky. In biotissues, in the visible range, the Rayleigh scattering occurs on striations in collagen, macromolecular aggregates, and membranes. The angular distribution of the scattered light for Rayleigh scattering is isotropic.

B.4.1.2 Mie Scattering

Mie scattering refers to light scattering by relatively large particles, which are of the order of the wavelength or larger. In biotissues, in the visible range, the Mie scattering occurs on cells, the nucleus, collagen fibers, and cellular components like mitochondria, lysosomes, etc.

The angular distribution of the Mie scattering is highly anisotropic. It depends strongly on the relative refractive index and size of the particle relative to the wavelength (the Mie-equivalent radius). For example, the greatest extinction of the forward scattered light is due to particles with dimensions between λ and 10λ. On the other hand, the particles with diameters between $\lambda/4$ and $\lambda/2$ are the dominant backscatters.

B.4.1.3 Scattering Anisotropy

During a scattering event, the electromagnetic wave changes its direction. The angular distribution of the scattering waves can be characterized by a scattering phase function $p(s,s')$, which is the probability density function for scattering in the direction s' of a photon traveling initially in the direction s. The scattering phase function can be determined experimentally from goniophotometric measurements in relatively thin tissue samples.[22] It also can be obtained theoretically/numerically. In particular, using Mie theory, the explicit expression for the scattering phase function can be obtained for spheroids. For other shapes (like ellipsoids and cylinders), the Scattering Phase Function (SPF) can be calculated numerically.

If scattering is symmetric relative to the direction of the incident wave, then the phase function depends only on the scattering angle θ: $p(\vec{s}, \vec{s}') = p(\vartheta)$. In particular, it is valid for spherical scatterers. However, more complex shapes (like ellipsoids and cylinders) can be oriented randomly with respect to the incident wave in the media, which breaks this symmetry. Nevertheless, multiple scattering acts on multiple scattering centers, distributed randomly in the biotissues, 'average' multiple individual scattering phase functions. Thus, the collective scattering phase function becomes symmetrical with respect to the incident wave direction and can be characterized by a single parameter: the scattering angle θ.

In biooptics, the most commonly used scattering phase function is Heyney–Greenstein SPF (HGPF):

$$p(\vartheta) = \frac{1}{4\pi} \frac{1-g^2}{(1+g^2 - 2g\cos\vartheta)^{3/2}} \tag{B.3}$$

Here, g is the scattering anisotropy parameter: $g = <\cos\theta>$. In particular, $g=0$ for isotropic (Rayleigh) scattering, $g=1-$ for total forward, and $g=-1-$ for total backward scattering.

For realistic biotissues, the scattering anisotropy g is 0.7 or more. Thus, it is primarily forward scattering. However, some tissue components can have much higher values. For example, blood has $g=0.998$.

In addition, other scattering phase functions are used to model the light propagation in tissues, including explicit Mie expressions[23] and a two-parameter Gegenbauer Kernel Phase Function (GKPF), which is a generalization of HGPF.[24]

B.4.1.4 Reduced Coefficient of Scattering

As we just mentioned, the scattering in biotissues is predominantly forward-bound. Thus, the photon retains some information about its initial direction after the individual scattering. Therefore, it can be helpful to assess the distance at which the photon 'forgets' its initial direction. Photon migration in turbid media can be considered a random walk at this scale.

This distance, l_s, is connected with the distance between individual scattering events L_s (here, as we already defined $L_s = 1/\mu_s$): $l_s = L_s / (1 - g)$. Thus, we can introduce the reduced coefficient of scattering:

$$\mu_s' = \mu_s(1-g) \tag{B.4}$$

which can be associated with isotropic photon migration in the turbid media.

B.4.1.5 Origins of Scattering in Turbid Tissues

Biological tissues are composed of various structures and levels of organization, such as cells, fibers, and macrostructures like blood vessels. These structures vary in size from a few tens of nanometers to hundreds of micrometers. Mammalian cells, for example, have diameters in the range of 5–75 micrometers. The epidermal layer of the skin is composed of large, uniform-sized cells. In contrast, fat cells can vary widely in size, from a few micrometers to 50–75 micrometers in normal cases and 100–200 micrometers in pathological conditions. Within cells, there are a variety of structures that affect how the tissue scatters light, such as cell nuclei (5–10 micrometers in diameter), mitochondria, lysosomes, and peroxisomes (1–2 micrometers), ribosomes (20 nanometers) and structures within organelles that can be up to a few hundred nanometers.

Scattering arises from mismatches in the refractive index of the components that make up the cell, which includes connective tissue fibers, cytoplasmic organelles, cell nuclei, and melanin granules. These components appear to blend at a microscopic level, with variations in the refractive index. In particular, the refractive index ranges from 1.35–1.36 (extracellular fluid) to 1.360–1.375 (cytoplasm) to 1.38–1.41 (nucleus, mitochondria, and organelles) to 1.46 (cell membrane) to 1.6–1.7 (melanin).[22]

As a result of this complex organization, in realistic tissues, both Rayleigh and Mie scattering occurs. Typically, Mie scattering is dominant in the range of 500nm and longer. However, in the range shorter than 500nm, the contribution of the Rayleigh scattering can be pretty significant.

B.4.2 Inelastic (Raman) Scattering

While elastic scattering is the predominant mechanism of electromagnetic wave scattering in turbid media, some photons may exchange energy with a scatterer (a molecule in this case) and excite the molecule's vibration/rotational vibration states. In this type of scattering, called inelastic or Raman scattering, the energy is transferred between the photon and the scattering center. As a result, the scattered photons usually have lower energy than the incident photons (Stokes shift). However, in certain circumstances, the scattered photons may have higher energy than the incident photons (anti-Stokes shift). By convention, the change in the photon energy in Raman spectroscopy (Stock shift) is characterized by a wavenumber shift, $\Delta\tilde{\nu}$, measured in [cm^{-1}]. The lifetime of Raman scattering, τ_R, is less than 1×10^{-14} s.

Inelastic scattering has a resonant origin, with peaks corresponding to the transitions between vibration/rotational–vibration states of a particular molecule. This combination of peaks is quite unique ('fingerprint') and can be used to identify molecules with high specificity. For biological samples, approximately 90% of the peaks are found in the so-called 'fingerprint' spectral region, covering ($\Delta\tilde{\nu} \sim 500$ cm^{-1} to $\sim 1,800$ cm^{-1}), with the remaining found in the higher energy CH/OH stretching vibrational modes covering ($\Delta\tilde{\nu} \sim 2,700$ cm^{-1} to $\sim 3,300$ cm^{-1}).[25]

In general, inelastic scattering is a rare event that happens at the rate of 10^{-8} of the elastic scattering.[26] In addition, the Stock shift for Raman scattering is relatively small. Thus, spontaneous Raman scattering can generally be observed in lab conditions with laser sources and specialized optics. However, multiple techniques can increase the Raman signal strength, including Surface-Enhanced Raman Scattering (SERS),[27] Tip-Enhanced Raman Scattering (TERS),[28] Stimulated Raman Scattering (SRS),[29] and Coherent Anti-Stokes Raman Scattering (CARS).[30] Raman enhancement techniques like SERS allow significant (on a scale of several orders of magnitude) increases in signal strength and SNR, thus increasing the method's sensitivity. Combined with very high specificity, it makes Raman spectroscopy a unique tool for many applications in biomedicine and beyond.

It should be pointed out that unlike fluorescence, which is sensitive to excitation light wavelength in a resonant manner (although usually, they are pretty broad peaks), the Raman scattering can be excited by light in a wide range of wavelengths as its signal intensity is inversely proportional to the wavelength of the excitation light.

Several practical considerations are commonly used during Raman spectroscopy. While shorter wavelengths (blue visible range and UV) generate a stronger signal, they also excite Autofluorescence of the tissue, which may mask the weak Raman signal. Consequently, the longer range of visible and NIR (500–830nm) is typically used, with a 785nm laser diode as the most common source, which balances the competing factors between Raman signal intensity, fluorescence, detector sensitivity, and cost. However, for samples with strong Autofluorescence, such as dyes, a 1,064 nm laser may be needed. For other applications, visible lasers in blue and green ranges (e.g., 532 nm) are becoming more common in Raman spectroscopy.

B.4.3 Laser Doppler

Dynamic (quasi-elastic) light scattering occurs when a moving object causes a shift in the frequency of the scattered light compared to the frequency of the light that was initially emitted. This shift is known as the Doppler effect, which is a change in the frequency of a wave caused by a change in the distance between the source of the wave and the receiver.

B.5 INTERFERENCE

According to the principle of superposition of waves, when two or more propagating waves of the same kind converge on a particular point, the resulting amplitude at that point equals the combined vector sum of the individual wave amplitudes. Interference occurs when two waves combine by adding their displacement at each point in space and time, resulting in a resultant wave with the greater, lower, or same amplitude. Constructive and destructive interference can arise from correlated or coherent waves that have the same or nearly the same frequency and originate from the same source. When the phase difference between the waves is an even multiple of π (180°), constructive interference occurs, while destructive interference arises when the phase difference is an odd multiple of π. If the phase difference lies between these two extremes, then the magnitude of the summed wave's displacement falls between its minimum and maximum values. Since light waves' frequency (~10^{14} Hz) is beyond the detectable range of currently available detectors, only the intensity of an optical interference pattern, which is proportional to the square of the average wave amplitude, can be observed.

B.5.1 Speckle Formation in Biotissues

Speckle structures are produced as a result of interference of a large number of elementary waves with random phases that arise when coherent light is reflected from a rough surface or when coherent light passes through a scattering medium. The speckle pattern remains static if the surface or media does not change. However, motion gives rise to spatial or temporal intensity fluctuations depending on the movement's velocity.[31] Thus, velocity estimation is performed by studying the temporal statistics of these fluctuations. A high-speed camera (~200fps) is used to measure and analyze the temporal statistics of the speckle patterns to estimate blood perfusion.

B.5.2 OCT

Optical Coherence Tomography (OCT) is founded on the principle of low-coherence interferometry, which involves directing a light beam with low coherence (high bandwidth) onto the tissue being examined and then combining the scattered, back-reflected light with a second beam (known as the reference beam), which is obtained by splitting the original light beam. The typical OCT setup includes a standard Michelson interferometer and a low time-coherence light source (e.g., a Superluminescent Diode or SLDs). The concept of the OCT setup is depicted in Figure B.3.

To retain coherence, OCT uses ballistic and near-ballistic photons.

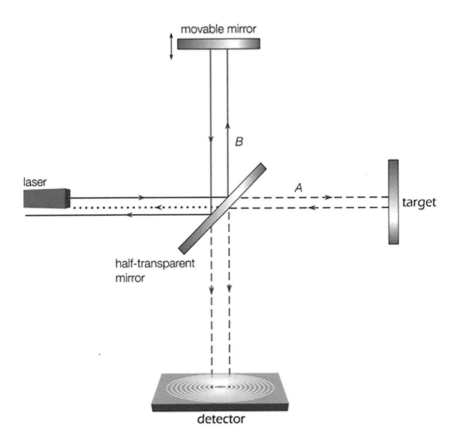

FIGURE B.3 The concept of the OCT setup.

There are many different implementations of OCT technology.[32] The two main OCT schemas are time-domain OCT (TD-OCT) and Fourier-domain OCT (FD-OCT). Fourier-domain OCT imaging can also be done in two ways: by spectral-domain OCT (SD-OCT) and swept-source OCT (SS-OCT).

Laterally adjacent depth scans (similar to the more familiar A-scans of ultrasound imaging technology) are used to obtain a sample's 2-D map of reflection sites. Note that, basically, there are two scan procedures in OCT: the OCT depth scan is performed by the reference mirror. The lateral OCT scan is performed by moving the sample or scanning the probe beam illuminating the specimen.

Direct acquisition of 2D OCT images is accomplished through another method, full-field OCT (FF-OCT).

B.6 OTHER TISSUE OPTICS CONCEPTS

B.6.1 Coefficient of Extinction

Equations B1 and B2 were introduced under the assumption of entirely non-scattering and non-absorbing media, respectively. However, in reality, media demonstrate both scattering and absorption. To accommodate for these phenomena, the coefficient of extinction can be introduced:

$$\mu_t = \mu_a + \mu_s \qquad (B.5)$$

Based on this definition, the propagation of collimated light in the medium with absorption and scattering can be described by the Beer-Lambert law:

$$I(l) = I_0 e^{-\mu_t l} \qquad (B.6)$$

The coefficient of extinction defines the Mean Free Path (MFP) or the length between two interactions: $l_{ph}=1/\mu_t$.

B.6.2 Light Propagation in Tissues

The light propagation in the tissue depends on the geometry and optical parameters of the tissue. Light propagation can generally be described by the Radiation Transfer Theory, or RTT, a Boltzmann equation for photons. The theory is valid for an ensemble of scatterers located far from one another. However, it is an integro-differential equation, which cannot be solved in the general case, particularly taking into account the multilayered structure and heterogeneity of the tissues. However, the equation can be simplified and solved in certain cases.

We can distinguish two important cases, $\mu_a \gg \mu_s$ and $\mu_s \gg \mu_a$, which are referred to as absorption-dominated and scattering-dominated regimes, respectively. For the absorption-dominated regime, just a small fraction of photons will return back in the reflectance mode, as most photons will be absorbed before being scattered back. This regime is not typical for biological tissues and may happen only in the strong absorption bands (e.g., the Soret band of hemoglobin or the 1,940nm water band). It is more typical for homogeneous media without strong scattering, e.g., liquids. The light intensity with depth in the absorption-dominated regime follows the Beer-Lambert law.

The scattering-dominated regime is a much more common scenario for biological tissues. In this case, the light is scattered multiple times between consecutive absorption events.

We can further consider several scenarios depending on several factors, including tissue thickness.

As the scattering in biological tissue in the optical range is predominantly forward (Mie scattering with $g > 0.7$), if the light undergoes just several scattering events, then the light path is close to a straight line. This scenario is referred to as ballistic photons. This scenario is particularly important for Optical Coherence Tomography or OCT. The photons that underwent just several scattering events can be gated using polarization optics, as photons gradually lose their polarization with every scattering event (retardation). Ballistic photons can also be gated using high numerical aperture optics.[33] In addition, the tissue reflectance for quasi-ballistic photons, the so-called single-backward scattering, can be calculated explicitly.[34]

If photons undergo multiple scattering events, eventually, they forget the initial direction, even in the case of strong forward scattering (g is close to 1). In this case, the photon path resembles a random walk, and such photons are referred to as diffuse photons. In this case, light propagation can be described by the diffusion equation. Consequently, this regime is called diffusion approximation. The diffusion equation can be solved for several important geometries with practical applications: a semispace, a slab of a certain thickness, and a segment between two concentric spheres with radii r_1 and r_2, which is important for hollow organs. However, diffusion approximation has several limitations. In particular, (a) mean optical free path ($1/\mu_t$) is much smaller than the size under consideration (e.g., source-probe distance, a distance from a light source or boundaries, and (b) $\mu_a \ll \mu'_s$ - photon has to be scattered multiple times before it is absorbed or exits the media.

B.6.3 Sampling Depth

Optical diagnostic methodologies rely on the fact that light propagating through the tissue collects information about optical tissue properties, which can be further deciphered into physiological or diagnostic information. As tissues typically have a multilayered heterogeneous structure, it is important to understand which part of the tissue was sampled.

We will briefly discuss it for typical geometries for optical diagnostics modalities: reflectance or transmittance geometries.

In transmittance geometry, the source and detector are placed on different sides of the tissue. Given that restrictions, transmittance geometry applications are typically limited to protruding body parts, like the finger or earlobe. However, it is the most widely used clinical modality, as pulse oximetry is a transmittance-based modality. In addition, the transmittance mode can also be used for larger body parts, like the head, in functional Near Infra Red Spectroscopy (fNIRS).

For transmittance-based modalities, the light collects information from all tissue layers between the source and detector. However, the optical path can deviate from a straight line as light experiences scattering. For example, for a pulse oximeter, the light passes through the nail blade, nail bed, muscle, phalanx bone, adipose tissue, dermis, and epidermis of the plantar side of the hand. In this case, the light experiences significant scattering in the epidermis, dermis, and phalanx bone. In general, simple rules can be deduced to the following: more scattering leads to a longer Mean Optical Path (MOP). Similarly, stronger absorption shortens the Mean Optical Path, as ballistic and quasi-ballistic paths have higher contributions.

For reflectance-based geometry, the light source and detector are placed on the same side of a body part. As it does not restrict the body part thickness, it can be applied virtually to any patch of skin or mucosa.

For reflectance-based geometry, there are two important cases to consider. In the first case, the source and detector are separated in space (e.g., spactally-resolved spectroscopy). It is the most common case for contact modalities, like Photoplesythmography (PPG). In this case, the light enters the tissue typically as a narrow beam (e.g., through an optical fiber). Similarly, the detector (or detectors) collects light from small skin patches located at predetermined distances from the sources. In this case, the light path between the source and detector has a characteristic 'banana' shape. The light propagation can be characterized by the Mean Optical Path

(MOP) and the Mean sampling Depth (MD). Chatterjee et al.[35] found that in reflectance mode pulse oximetry (660 nm and 940nm wavelengths), the Mean Optical Path (MOP) and the Mean sampling Depth (MD) depend almost linearly on the source-detector separation. MOP and MD for Red and IR channels are identical until 6mm and 5mm, respectively. After that, red and IR paths start diverging. Thus, for space-separated geometries, MOP and MD are determined primarily by geometrical factors, namely the source-detector distance, d. In particular, the sampling depth for this configuration is proportional to $\sqrt{2}d/4$ in the weak absorption limit ($d<<\delta$) and $\sqrt{d\delta}/2$ in the strong absorption limit ($d>>\delta$).[36] Here δ is the light penetration depth. Thus, these types of probes are designed to probe a particular depth/layer (or layers) of the tissue.

In the second case, the source-detector distance is not important. It is, for example, typical for imaging applications, where the tissue is illuminated by a broad beam (or ambient light), and the data are collected from a relatively large skin area. In this case, the light propagation can also be characterized by the Mean Optical Path (MOP) and Mean sampling Depth (MD). However, unlike space-separated geometry, they are determined by the optical properties of the tissue. For example, in this case, the sampling depth is close to the light penetration depth δ. MOP can be estimated as 7.8δ.[37] As such, the sampling depth may vary for different wavelengths. Similarly to the transmittance mode, simple rules can be deduced: more scattering leads to a longer Mean Optical Path, and stronger absorption shortens the Mean Optical Path, as ballistic and quasi-ballistic paths have higher contributions.

There are several concepts related to sampling depth.

Penetration depth measures how deep light or any electromagnetic radiation can penetrate a media. The penetration depth is defined as the depth where light intensity falls to the level $1/e$ of its initial power. The penetration depth depends not just on the wavelength but also on the type of illumination. If the tissue is illuminated with a collimated light source (e.g., laser beam or optical fiber), then the incident beam intensity in the tissue follows the Beer-Lambert law. Namely, it decays exponentially with distance z as $exp(-\mu_t z)$. Thus, the penetration depth for collimated light will be $\delta=1/\mu_t$. However, the light intensity of the diffuse light will follow the exponential law with $exp(-\mu_{eff} z)$. Here $\mu_{eff} = \sqrt{3\mu_a(\mu_a + \mu_s')}$. The same applies to diffuse illumination (e.g., ambient light). Thus, the penetration depth for diffuse light will be $\delta=1/\mu_{eff}$.

Optical Density or OD. Optical Density is a standard measure of optical transparency for optical filters, but it is occasionally used to characterize tissue transparency. Optical Density can be defined as $OD=\mu_t d$, where d is the thickness of the layer. However, typically Optical Density is defined as $log_{10}(I_{in}/I_{out})$, where I_{in} and I_{out} are intensities of incoming and outcoming beams, respectively (note that these definitions differ only by the factor of $ln10$). Thus, OD4 means that the filter's transmission is 10^{-4}. Optical Density is broadly used in microbiology to characterize bacterial concentrations in inoculums.

In addition, for imaging geometries, we can define the maximum depth of visibility of a certain defect (e.g., capillary in capillaroscopy). This depth can be linked to a contrast ratio resolved by the optical system. Saiko et al.[38] showed that the contrast ratio of the defect buried on the depth z follows the exponential law with $exp(-2\mu_{eff} z)$.

B.6.4 Reflectance

Tissue reflectance is the ratio of the reemerged light to the incident light. The total tissue reflectance consists of two major components: specular reflectance from the surface of the tissue (see Ray reflection section) and diffuse reflectance. Diffuse reflectance contains photons, which travel through the bulk of the tissue.

The tissue reflectance varies dramatically in the optical range of the spectrum.

UV range: the amount of light reflected back from the epidermis is minimal in the Ultraviolet range for wavelengths shorter than 300 nm. The depth of light penetration in the epidermis is limited to a few cell layers, and the epidermal chromophores have a limited effect on the diffuse reflectance spectrum.

The amount of radiation bouncing back from the skin is higher than predicted by Fresnel's law (specular reflection) in the UVA range (315–400 nm). This type of radiation can penetrate the epidermis to depths of hundreds of micrometers, and the skin's colors can alter the reflectance spectrum's shape.

Visible range: when using visible light within the 400–800 nm range, the penetration depth into the tissue is between 0.5 mm and 2.5 mm. In this range, the diffuse reflectance spectrum is mainly influenced by absorption and scattering. As the light penetrates deeper into the tissue, the fraction of backscattered light increases due to multiple scattering within the skin. As a result, the diffuse reflectance (Rd) ranges from 15% to 70%. The reflectance spectrum has a distinct minimum in the range of 415–430 nm due to hemoglobin absorption in the dermis (Soret band). The reflectance spectrum also has characteristic dips in the range of 540–580 nm due to the Q-absorption bands of hemoglobin. Additionally, weak minima in the reflectance may be observed due to the absorption of carotene (480 nm) and bilirubin (460 nm).

Infrared range: when using light within the 800–1,500 nm range, absorption is even lower, scattering dominates, and the penetration depth is 3.5 mm and more. The light is entirely diffuse, resulting in an increased diffuse reflectance of 35–70%. In the Near-Infrared spectral range, the skin reflectance increases until 800–900 nm and then decreases due to the absorption bands of water at 970 nm, 1,190 nm, and 1,450nm.

B.6.5 Therapeutic Windows

Certain wavelength ranges in NIR with large penetration depth are known as optical or biological windows. The boundaries of these ranges are defined by water absorption, which is prominent in the NIR range.

The first optical window ranges between 650 nm and 950 nm. Due to reduced absorption, it allows for deeper depth penetration in tissue than in the visible region.

A second NIR optical window contains wavelengths from 1,100 nm to 1,350 nm. It is used in multiple imaging and therapeutic applications. The first and second windows are also known as therapeutic windows, as they are used in numerous therapeutic applications.

Two additional NIR optical windows were proposed recently for imaging applications. A third therapeutic spectral window with wavelengths between 1,600 nm and 1,870 nm[39] was proposed, as this range can be used for imaging more deeply into the tissue due to a reduction in scattering.

The 4th NIR window, 2,100–2,300 nm, was recently suggested for imaging biomedical applications.[40] However, using this range is in its infancy, as no suitable probes exist.

B.7 THERMOGRAPHY

Unlike contact temperature measurement techniques (e.g., mercury thermometer), noncontact (remote) ones measure not the temperature directly, but rather energy flow from the object and derive temperature from these measurements (radiometry).Radiometry refers to the measurement of electromagnetic radiation emitted by an object and is based on the fact that all objects with temperatures above 0K emit electromagnetic radiation. Most radiation emitted by objects in the temperature range 0–100° is in the thermal infrared range, specifically wavelengths three micrometers and longer. For example, the peak radiation from an object at a temperature of 300K (27°) is at 9 micrometers. Objects in this temperature range do not emit noticeable radiation in the visible range of the spectrum. Only objects heated to 500° and above emit detectable radiation in the visible range of the spectrum.

Radiometry is based on the concept of a blackbody, which is an idealized object that absorbs all incident electromagnetic radiation and emits all of the absorbed energy, with an emissivity of 1.0 or 100%. The emissivity of an object is the ratio of its energy flow to that of a blackbody at the same temperature. As a result, a 2-D imaging array of thermal IR sensors does not directly measure temperature but instead measures the energy flow emitted by the target. IR sensors detect energy flow over a wide spectrum range, and temperature is calculated based on these measurements. The amount of emitted radiation depends on both the material's temperature and its emissivity.

Indirect thermal measurements occur in the atmosphere, and the absorption of atmospheric gases may impact the thermal energy flow. Noticeable absorbers in the three micrometers and above spectral range are carbon dioxide and water vapor. While carbon dioxide has a narrow absorption peak of around 4.3μm, water has a broad and strong absorption peak in the range of 5–8μm. This absorption peak separates the thermal radiation range into two spectral ranges or atmospheric transparency windows, which are used in practical applications: the middle (MWIR, 3–5μm) and long (LWIR, 8–14 μm) wavelength Infrared (IR) spectral ranges.

Thermography measures the energy flow emitted by the body's surface (e.g. skin). Thus, it measures the surface (skin) temperature, which is the balance between heat generation and dissipation. The skin, which covers an area of approximately 2 sq. m, is the body's largest organ and plays a crucial role in regulating body temperature. In conjunction with adipose tissue, it acts as an insulator to keep the body warm. It also functions as a 'heat radiator' system, responsible for about 90% of heat loss from the body. The emissivity of human skin is close to that of a perfect blackbody, with a value of at least 0.91 in the medium-wave infrared range and even higher (0.97–0.98) in the long-wave infrared spectrum. Because of this high emissivity, it is well-suited for thermographic assessment, with minimal reflected energy flow affecting measurements. However, it is still recommended to avoid measuring temperature near hot objects.

NOTES

1. S. Prahl, http://omlc.ogi.edu/spectra/. 2001.
2. A Roggan, M Friebel, K Dörschel, et al. "Optical properties of circulating human blood in the wavelength range 400–2500 nm", *J Biomed Opt*, 4: 36–46 (1999).
3. MR Schaffer, U Tantry, PA Efron, et al. "Diabetes-impaired healing and reduced wound nitric oxide synthesis: A possible pathophysiologic correlation", *Surgery*, 121: 513–519 (1997).
4. M Minetti, W Malorni, "Redox control of red blood cell biology: The red blood cell as a target and source of prooxidant species", *Antioxid Redox Signal*, 8: 1165–1169 (2006).
5. WAP Luck Infrared overtone region. In WAP Luck (editor), *Structure of Water and Aqueous Solutions*, Verlag Chemie, Weinheim, 1974, pp. 248–284.
6. GM Hale, MR Querry, "Optical constants of water in the 200nm to 200 micron wavelength region", *Appl Opt*, 12: 555–563 (1973).
7. HQ Woodard, DR White, "The composition of body tissues", *Br J Radiol*, 59: 1209–1218 (1986).
8. R Nachabe, JW van der Hoorn, R van de Molengraaf, et al. "Validation of interventional fiber optic spectroscopy with MR Spectroscopy, MAS-NMR Spectroscopy, high-performance thin-layer chromatography, and histopathology for accurate hepatic fat quantification", *Invest Radiol*, 47: 209–216 (2012).
9. CE Cooper, SJ Matcher, JS Wyatt, et al. "Near-infrared spectroscopy of the brain: Relevance to cytochrome oxidase bioenergetics", *Biochem Soc Trans*, 22: 974–980 (1994).
10. JT Eells, MTT Wong-Riley, J VerHoeve et al. "Mitochondrial signal transduction in accelerated wound and retinal healing by near-infrared light therapy", *Mitochondrion*, 4: 559–567 (2004).
11. JM Lunetta, RA Zulim, SR Dueker, et al. "Method for the simultaneous determination of retinol and β-carotene concentrations in human tissues and plasma", *Anal Biochem*, 304: 100–109 (2002).
12. ME Darvin, I Gersonde, M Meinke, et al. "Non-invasive in vivo determination of the carotenoids beta-carotene and lycopene concentrations in the human skin using the Raman spectroscopic method", *J Phys D: Appl Phys*, 38: 26962700 (2005).
13. ME Darvin, I Gersonde, H Albrecht, et al. "Determination of beta carotene and lycopene concentrations in human skin using resonance Raman spectroscopy", *Laser Phys*, 15:295–299 (2005).
14. MJ Leahy, JG Enfield, NT Clancy, et al. "Biophotonic methods in microcirculation imaging", *Med Laser Appl*, 22: 105–126 (2007).
15. J Xia, J Yao, LV Wang, "Photoacoustic tomography: Principles and advances", *Electromagn Waves (Camb)*, 147:1–22 (2014).
16. LV Wang, "Tutorial on photoacoustic microscopy and computed tomography", *Selected Topics in Quantum Electronics, IEEE J*, 14:171–179 (2008).

17. P Beard, "Biomedical photoacoustic imaging", *Interface Focus*, 1: 602–631 (2011).
18. T Burton, G Saiko, A Douplik, "Towards development of specular reflection vascular imaging", *Sensors*, 22: 2830 (2022).
19. WM Star, "Diffusion theory of light transport", in AJ Welch (editor), *Optical-Thermal Response of Laser-Irradiated Tissue* (2nd Ed), Springer, Dordrecht, NLD, 2011, pp. 145–202.
20. AN Bashkatov, EA Genina, VV Tuchin, "Optical properties of skin, subcutaneous, and muscle tissues: A review", *J Innov Opt Health Sci*, 4(1): 9–38 (2011).
21. H Ding, JQ Lu, WA Wooden, et al. "Refractive indices of human skin tissues at eight wavelengths and estimated dispersion relations between 300 and 1600 nm", *Phys Med Biol*, 51: 1479–1489 (2006).
22. VV Tuchin, *Tissue Optics: Light Scattering Methods and Instruments for Medical Diagnosis*, SPIE Press, Bellingham, WA, 2007.
23. CF Bohren, DR Huffman, *Absorption and Scattering of Light by Small Particles*, Wiley, New York, 1983.
24. PW Barber, SC Hill, *Light Scattering by Particles: Computational Methods*, World Scientific, Singapore, 1990.
25. CH Camp Jr, MT Cicerone, "Chemically sensitive bioimaging with coherent Raman scattering", *Nat Photonics*, 9:295–305 (2015).
26. HJ Bowley, DJ Gardiner, DL Gerrard, et al. *Practical Raman Spectroscopy*, Springer, Berlin Heidelberg, 1989.
27. M Fleischmann, PJ Hendra, AJ McQuillan, "Raman spectra of pyridine adsorbed at a silver electrode", *Chem Phys Lett*, 26:163–166 (1974).
28. MD Sonntag, JM Klingsporn, LK Garibay, et al. "Single-molecule tip-enhanced Raman spectroscopy", *J Phys Chem C*, 116:478–483 (2011).
29. MBJ Roeffaers, X Zhang, CW Freudiger, et al. "Label-free imaging of biomolecules in food products using stimulated Raman microscopy", *J Biomed Opt*, 16:021118 (2011).
30. J-X Cheng, XS Xie, "Coherent anti-Stokes Raman scattering microscopy: Instrumentation, theory, and applications", *J Phys Chem B*, 108:827–840 (2004).
31. JD Briers, "Laser speckle contrast imaging for measuring blood flow", *Opt Appl*, 37: 139–152 (2007).
32. DP Popescu, LP Choo-Smith, C Flueraru, et al. "Optical coherence tomography: Fundamental principles, instrumental designs and biomedical applications", *Biophys Rev*, 3(3): 155 (2011).
33. A Pandya, I Schelkanova, A Douplik, "Spatio-angular filter (SAF) imaging device for deep interrogation of scattering media", *Biomed Opt Express*, 10(9):4656 (2019).
34. G Saiko, A Douplik, "Reflectance of biological turbid tissues under wide area illumination: Single backward scattering approach", *Int J Photoenergy*, 241364 (2014).

35. S Chatterjee, PA Kyriacou, "Monte Carlo analysis of optical interactions in reflectance and transmittance finger photoplethysmography", *Sensors*, 19: 789 (2019).
36. S Feng, F Zeng, B Chance, "Monte Carlo simulations of photon migration path distributions in multiple scattering media", *Proc SPIE*, 1888: 78–97 (1993).
37. S Jacques, I Saidi, A Ladner, D Oelberg, "Developing an optical fiber reflectance spectrometer to monitor bilirubinemia in neonates", *Proc SPIE*, 2975: 115–124 (1997).
38. G Saiko, A Douplik, "Visibility of capillaries in turbid tissues: an analytical approach", arXiv:2210.04301 (2022).
39. LA Sordillo, S Pratavieira, Y Pu, et al. "Third therapeutic spectral window for deep tissue imaging", *Proc SPIE* 8940, Optical Biopsy XII, 89400V (2014).
40. LA Sordillo, Y Pu, S Pratavieira, et al. "Deep optical imaging of tissue using the second and third near-infrared spectral windows", *J Biomed Opt*, 19:056004 (2014).

APPENDIX C

Physiological Imaging

Design Considerations

WITH THE PROLIFERATION OF smartphones, healthcare is on the verge of transformational changes. The widespread use of smartphones has the potential to decrease costs, democratize healthcare, and bring diagnostic capabilities closer to patients.

One particular aspect of these transformational changes is the emergence of multiple novel medical imaging technologies. This transformation started with applications of traditional digital imaging, sometimes called anatomical imaging in this context. For example, anatomical imaging is used in wound management to track time progress[1] and dermatological applications, e.g., skin cancer diagnostics.[2]

In addition to anatomical imaging, inexpensive CMOS cameras enabled imaging modalities, which extract underlying physiological information. This group of technologies can be called physiological imaging based on this intent. An example of such modalities can be Photoplethysmographic (PPG) imaging for extraction of heart rate,[3] hyperspectral/multispectral imaging for extraction of tissue blood oxygenation,[4] or fluorescence imaging of bacterial load.[5] In addition to these mainstream applications, physiological imaging can monitor capillary grid[6] or extract pulse wave velocity.[7]

The significant subset of these technologies (other than thermographic and fluorescence imaging) is based on tissue (skin or mucosa) reflectance. In this case, physiological imaging typically samples the signal from

1.5 mm to 2mm depth.[8] Thus, the signal can be pretty faint, and its robust collection may represent some challenges.

In addition, as all physiological imaging technologies are still in the early stages of clinical translation (experimental modalities or early clinical adopters), not a large body of evidence has yet been accumulated. Thus, data quality and reproducibility issues have not emerged yet. However, this topic may get prominence with data accumulation as it happened with other optical modalities. For example, it is known that the Laser Doppler Flowmetry (LDF) exhibits significant spatial and temporal variations in LDF measurements.[9]

This Appendix aims to provide background information on data collection from technical and physiological points of view and identify issues affecting data quality and reproducibility during physiological imaging. We will follow[10] closely in this discussion.

The approach can be applied to various camera types, including CMOS and charge-coupled device (CCD). However, we will illustrate it mainly using CMOS examples.

C.1 FACTORS

The primary purpose of virtually any physiological imaging technique is to extract and quantify a physiologically relevant parameter: heart rate, tissue blood saturation, etc. Thus, as the first step, we need to establish a reference point for the analytical ability of a method. Then, we need to link it to the camera properties. Finally, we need to understand the requirements for the resolution of the system.

C.1.1 Limits of Detect and Quantification

In analytical methods (including clinical laboratory tests), there are two relevant levels of signal quantification: Limit of Detection (LOD) and Limit of Quantification (LOQ) (see, for example,[11]).

The Limit of Detection is defined as an analyte content that can be distinguished from the blank (no analyte) with an error probability of $(1-\beta)$. Here β is the likelihood of false-negative results; Type II error.

The Limit of Quantification is defined as an analyte content that can be determined with a certain (often arbitrary) level of precision.

LOD and LOQ can be defined using the mean and variance of the signal, which can be combined in a single number metric, Signal-to-Noise Ratio (SNR). Typically, LOD and LOQ are defined using SNR with values of 3 (or 3.3) and 10 for LOD and LOQ, respectively (see, for example,[12]).

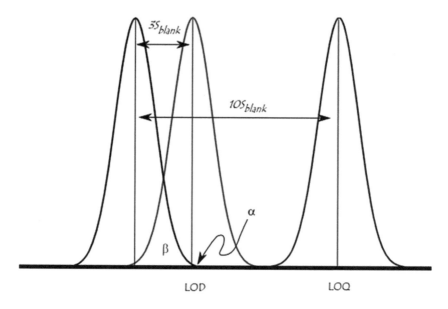

FIGURE C.1 Illustration of the concept of detection limit and quantitation limit by showing distributions associated with blank, detection limit (LOD), and quantitation limit (LOQ) level samples. (Reproduced from under CC BY-SA 4.0 license). https://en.wikipedia.org/wiki/Detection_limit.

Figure C.1 illustrates the relationship between the blank, the Limit of Detection (LOD), and the Limit of Quantitation (LOQ) by showing the probability density function for normally distributed measurements at the blank, at the LOD defined as 3 × standard deviation of the blank, and at the LOQ defined as 10 × standard deviation of the blank. For a signal at the LOD, the alpha error (probability of false-positive) is small (1%). However, the β error (probability of a false-negative) is 50% for a sample that has a concentration at the LOD. This means a sample could contain an impurity at the LOD, but there is a 50% chance that a measurement would give a result less than the LOD. At the LOQ, there is minimal chance of a false-negative.

As we are interested in quantitative measurements, we will use LOQ with SNR=10 in further calculations.

C.1.2 CMOS Camera Noise

To assess SNR for imaging technology, we need to identify the mean and variability of the signal. Some variability of the signal can be attributed to noises. In particular, the CMOS camera has several types of noise: shot noise, dark noise, and quantization noise.

Dark noise is a statistical variation in the number of electrons thermally generated within the pixel in a photon-independent fashion and is the electron equivalent of photon shot noise. Dark noise is a statistical variation, so it cannot be avoided by applying background subtraction. The background subtraction is applied to remove the average of dark electron, that is, 'dark current'.

Dark current, and therefore dark noise are temperature-dependent, with less noise at lower temperatures. For most biological experiments, dark current and dark noise are negligible over a typical exposure interval of less than five minutes. Because dark noise is typically negligible, the main noise component coming from the camera that needs to be considered is read noise.

The shot noise can be modeled by a Poisson process. In this case, the variance of the noise σ_p^2 is equal to the mean number of photons μ_p. The dark noise consists of read noise and dark current. However, the dark noise can be ignored in most practical cases other than low light conditions (e.g. astronomy). The quantization noise arises from digitizing the continuous voltage signal into a digital one and can be modeled by a Gaussian distribution. The read noise, dark current, and quantization noise are set values for a particular camera and typically can be found in spec sheets.

We can take into account that sources of noise are independent and use error propagation rules to write expressions for the camera output mean and SNR:

$$\mu = K\left(\mu_d + \eta\mu_p\right) \quad \text{(C.1a)}$$

$$SNR = \frac{\eta\mu_p}{\sqrt{\sigma_d^2 + \sigma_q^2/K^2 + \eta\mu_p}} \quad \text{(C.1b)}$$

Here K is the camera's sensitivity; η is the sensor's quantum efficiency (wavelength dependent), μ and σ refer to mean and standard deviations for photons (subindex p), dark noise (subindex d), quantization (subindex q), and output (no subindex).

However, for simplicity, we will ignore dark noise and quantization noise. This assumption is generally valid for all cases other than low-lit conditions. In this case, the SNR can be written as (here subscript 1 refers to the single-pixel SNR):

$$SNR_1 = \frac{\eta\mu_p}{\sqrt{\eta\mu_p}} = \sqrt{\eta\mu_p} \qquad (C.2)$$

C.1.3 Sampling Frequency

The next important step is to understand the requirements for the resolution of the system. We can illustrate this in a specific example. One particular application of physiological imaging is determining the parameters of the capillary grid, which can be used for shock progression monitoring[13] or cancer transformation detection.

In most skin parts, the capillaries are arranged vertically (hairpins). Thus, from the surface, they can be perceived as dots. However, the contrast between these points is relatively low. Therefore, to detect them, we have to account for two factors: (a) the contrast ratio associated with them needs to be detectable by the sensor (pixel size ↑), and (b) the spatial sampling frequency should satisfy the Nyquist–Shannon sampling theorem (pixel size ↓).

C.2 SYSTEM DESIGN ASPECTS

This section will illustrate how to incorporate physiological considerations into the imaging system design.

We can consider it a three-step process. The first step is to understand the minimal target size that must be resolved. This knowledge will help determine the spatial sampling frequency and, ultimately, pixel sampling area size (pixel size). The next step is determining the number of photons required to get an adequate signal. Finally, the pixel sampling area can be linked with the sensing area on the sensor (single or several sensor cells binned together) using the optical system.

In our narrative, we will mainly use the pixel size to annotate the size of the target area, which is sampled by a camera pixel.

C.2.1 Sampling Tissue Heterogeneities

Each type of physiological measurement comes with its scale. In our capillaries visualization example, it is the distance between individual capillaries.

The distance between capillaries can be estimated from the mean capillary density. For example, Tibirica et al.[14] found that in normal individuals, the mean capillary density ranges from 75mm^{-2} in feet to 120mm^{-2} in hands. Consequently, we can estimate intercapillary distances as $n^{-1/2}$= 0.1mm.

Thus, according to the Nyquist–Shannon sampling theorem, the spatial sampling frequency should be at least two times higher than this value. Considering that this spatial sampling frequency is limited by the pixel's largest dimension (its diagonal), the maximal pixel size can be estimated at around 35µm.

C.2.2 Number of Photons Required

As we already mentioned, a realistic physiological signal can be pretty faint. For example, the amplitude of the remote PPG signal (see Figure C.2) can be on a scale of 1% of the average value (background). Thus, in this case, the useful signal is the modulation, not the background itself. To illustrate it, let's suppose that the mean of the background signal is μ. If we consider its slight change in time to $\mu'=\mu-\Delta$, (here μ and Δ are measured in pixel intensity units, which are proportional to the photon counts $\mu=K\eta\mu_p$) then the SNR for physiological modulation can be found as:

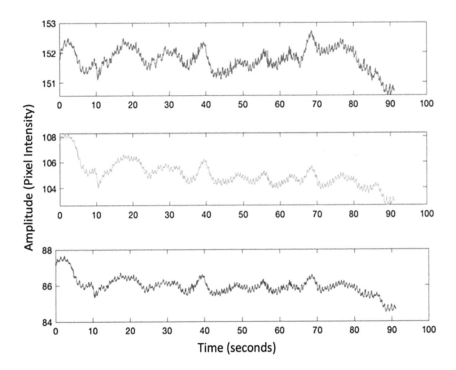

FIGURE C.2 An example of a remote PPG signal. Each subplot represents raw data collected by red (top panel), green (middle panel), and blue channels (bottom panel), respectively. Low-frequency (0.1 Hz) oscillations are visible in addition to the cardiac pulsations.

$$SNR_{phys} = \frac{\mu - \mu'}{K\sqrt{\eta\mu_p}} = \frac{\Delta}{K\sqrt{\eta\mu_p}} = \frac{\Delta}{\mu}\sqrt{\eta\mu_p} = \frac{\Delta}{\mu}SNR_{background} \quad (C.3)$$

Consequently, to quantify the physiological signal, we need to meet LOQ (namely, SNR at least 10:1) for the signal on the scale Δ/μ of the background ('blank' in analytical methods terminology).

Alternatively, instead of temporal variation (PPG signal), we can consider spatial changes in reflectance (e.g. capillary visualization). In this case, Δ/μ in Eq.C.3 can be substituted with the contrast ratio.

By reversing Eq.C.3, we can write an explicit expression for the number of photons which we need to detect per measurement:

$$\mu_p = \frac{1}{\eta}\left(\frac{\mu}{\Delta}LOQ\right)^2 \quad (C.4)$$

Here, the μ/Δ ratio is used for ballpark estimation. In our example ($\Delta/\mu=0.01$) using Eq.C.4, we can estimate that we need around 10^6 photons per measurement.

C.2.3 Design Considerations

Based on Eq.C.2, a simple observation can be derived: to increase SNR, we need to increase the number of photons detected.

Thus, we can either increase the density of photon flux near the detector (by changing distance and increasing illumination intensity) or increase the sensor size or integration time (e.g. spatial or temporal binning). These two approaches will be briefly discussed below.

Distance

Distance between the imaging system and the target area may impact the collected signal in several ways. Primarily, it affects the strength of the signal, which will be discussed further. However, it also impacts the sampling depth through spatio-angular gating.[15]

To analyze the impact of the distance on the signal strength, we can consider two imaging geometries: imaging the large skin area to collect averaged data (e.g. remote PPG or rPPG) and searching for small clusters within the target area (e.g. bacterial imaging or thermography).

In addition, we can split modalities into active and passive ones. This separation depends on whether the device emits any type of energy (active device) or not (passive device). For example, pertinent to imaging devices, if the device provides its own illumination, it can be considered active;

if not, it is passive. Examples of passive modalities can be rPPG, which relies on ambient light illumination, or Long-Wave Infrared (LWIR) thermography.

To analyze light propagation and its dependence on distance, we can split light propagation into three components:

- Incident light, which illuminates the target area (other than thermography);
- Light propagation and reflectance within the tissue (other than thermography);
- Light from the target area to the sensor.

Tissue propagation and reflection will not be impacted by the distance other than mentioned above spatio-angular gating. Thus, we can focus on the forward (imaging system -> target) and backward (target -> imaging system) light propagation.

To simplify calculations, we can make further assumptions. In particular, let's assume that the light source is close to the camera in an active modality. Thus, both distances – illumination system – target area and target area – sensor are equal to L. We also assume a simplified illumination model (a point source). However, this model is relevant for many clinical (particularly telehealth) smartphone-based applications.

Dependences of the sensor signal on the distance from the camera to the target area, L for the large (target is larger than a pixel size) and small (target is smaller than a pixel size) area imaging for the point light source are presented in Tables C.1 and C.2, respectively.

C.2.3.1 Binning

As we discussed earlier, in a realistic physiological imaging scenario, we need to collect photons on the scale of millions. This scenario can be unattainable for one Active Pixel Sensor (APS) cell. For example, the pixel's full-well capacity can be on the scale of 10,000–20,000 photons. It can be addressed using binning.

Binning is the grouping of outputs collected from several cells (pixels). It is an efficient way to increase sensitivity and SNR. In CCD devices, binning can be achieved on the sensor level. For CMOS devices, binning cannot be performed on the sensor level; however, it can be done digitally. It is less efficient than sensor-level binning[16] (for

TABLE C.1 Dependence of the sensor signal on the distance from the camera to the target area, L in the case of the large area imaging (target is larger than a pixel size)

	Active Modality	Passive Modality
Incident light intensity at the target for the point source	$\sim 1/L^2$	constant
The intensity of light on the sensor from the element of surface in the target area	$\sim 1/L^2$	$\sim 1/L^2$
The area sampled per pixel	$\sim L^2$	$\sim L^2$
Total effect	$\sim 1/L^2 * 1/L^2 * L^2 \sim 1/L^2$	$\sim 1/L^2 * L^2 \sim L^0$

example, it cannot reduce read noise); however, it provides significant SNR improvement anyway.

In most scenarios, binning is achieved by averaging signals over NxN cells. Taking into account that sources of noise in each cell are independent, we can write:

$$SNR_{NxN} = \frac{\sum_{NxN} \eta \mu_p}{\sqrt{\sum_{NxN} \eta \mu_p}} = \frac{N^2 \eta \mu_p}{\sqrt{N^2 \eta \mu_p}} = N\sqrt{\eta \mu_p} = N \times SNR_1 \quad (C5)$$

Thus, NxN binning allows approximately Nx improvement in SNR in the CMOS camera. However, it should be noted that binning reduces the image's resolution by the factor of N.

Another way to improve the SNR of the signal is to use temporal binning (image stacking). If N images are stacked together, then

$$SNR_N = \frac{\sum_N \eta \mu_p}{\sqrt{\sum_N \eta \mu_p}} = \frac{N \eta \mu_p}{\sqrt{N \eta \mu_p}} = \sqrt{N}\sqrt{\eta \mu_p} = \sqrt{N} \times SNR_1 \quad (C6)$$

Note that for temporal binning, there is no decrease in image resolution.

TABLE C.2 Dependence of the sensor signal on the distance from the camera to the target area, L in the case of the small area imaging (target is smaller than a pixel size)

	Active Modality	Passive Modality
Incident light intensity at the target for the point source	$\sim 1/L^2$	constant
The intensity of light on the sensor from the element of surface in the target area	$\sim 1/L^2$	$\sim 1/L^2$
The area sampled per pixel	$\sim L^0$	$\sim L^0$
Total effect	$\sim 1/L^2 * 1/L^2 * L^0 \sim 1/L^4$	$\sim 1/L^2 * L^0 \sim 1/L^2$

NOTES

1. C Wang, JAE Anderson, R Evans, et al. "Point-of-care wound visioning technology: Reproducibility and accuracy of a wound measurement app", Facchiano A, editor. *PLOS ONE*, 12(8):e0183139 (2017).
2. AP Kassianos, JD Emery, P Murchie, FM Walter, "Smartphone applications for melanoma detection by community, patient and generalist clinician users: A review", *Br J Dermatol*, 172(6):1507–1518 (2015).
3. AV Moço, S Stuijk, G de Haan, "New insights into the origin of remote PPG signals in visible light and infrared", *Sci Rep*, 8(1):8501 (2018).
4. G Lu, B Fei, "Medical hyperspectral imaging: A review", *J Biomed Opt*, 19(1):010901 (2014).
5. AR Oropallo, C Andersen, R Abdo, et al. "Guidelines for point-of-care fluorescence imaging for detection of wound bacterial burden based on delphi consensus", *Diagnostics*, 11(7):1219 (2021).
6. R Kanawade, G Saiko, A Douplik, "Monitoring of epithelial capillary density", InC Depeursinge, A Vitkin (editors),*Novel Optical Instrumentation for Biomedical Applications IV*,SPIE,2009. p. 73711L – 1. (SPIE-OSA Biomedical Optics; vol. 7371).
7. G Saiko, M Dervenis, A Douplik, "On the feasibility of pulse wave velocity imaging for remote assessment of physiological functions", *Adv Exp Med Biol*, 1269: 393–397 (2021).
8. AV Moço, S Stuijk, G de Haan, "New insights into the origin of remote PPG signals in visible light and infrared", *Sci Rep*, 8(1):8501 (2018).
9. T Tenland, EG Salerud, GE Nilsson, PA Oberg, "Spatial and temporal variations in human skin blood flow", *Int J Microcirc Clin Exp*, 2(2):81–90 (1983).
10. G Saiko, "Physiological imaging: Why pixel size matters", arXiv:2302.00404(2023).
11. Clinical and Laboratory Standards Institute, Protocols for determination of limits of detection and limits of quantitation, Approved Guideline. CLSI, 2004.
12. AS Lister, "Validation of HPLC methods in pharmaceutical analysis", In *Separation Science and Technology* [Internet], Elsevier, 2005 [cited 2022 Feb 14]. p. 191–217. Available from: https://linkinghub.elsevier.com/retrieve/pii/S0149639505800510.
13. G Saiko, A Pandya, I Schelkanova, et al. "Optical detection of a capillary grid spatial pattern in epithelium by spatially resolved diffuse reflectance probe: Monte Carlo verification", *IEEE J Sel Top Quantum Electron*, 20(2):187–95 (2014).
14. E Tibiriçá, E Rodrigues, R Cobas, MB Gomes, "Increased functional and structural skin capillary density in type 1 diabetes patients with vascular complications", *Diabetol Metab Syndr*, 1(1):24 (2009).
15. A Pandya, I Schelkanova, A Douplik, "Spatio-angular filter (SAF) imaging device for deep interrogation of scattering media", *Biomed Opt Express*, 10(9):4656 (2019).

16. SU Rehman, A Kumar, A Banerjee, "SNR improvement for hyperspectral application using frame and pixel binning", InXJ Xiong, SA Kuriakose, T Kimura (editors), New Delhi, India, 2016 [cited 2023 Jan 14]. p. 98810Y. Available from: http://proceedings.spiedigitallibrary.org/proceeding.aspx?doi=10.1117/12.2220599.

APPENDIX D

Artificial Intelligence

D.1 ARTIFICIAL INTELLIGENCE

Artificial Intelligence (AI) is a field of computer science that focuses on creating machine systems that can perform tasks that typically require human intelligence. AI aims to develop computer systems that can perceive and understand the environment, reason and make decisions, learn from experience, and interact with humans in a natural and intelligent manner. AI techniques and methodologies include machine learning, Natural Language Processing (NLP), computer vision, robotics, expert systems, knowledge representation, and reasoning, amongst others, to achieve these goals. These techniques enable AI systems to process and analyze vast amounts of data, understand human (natural) language, recognize patterns, make decisions, solve complex problems, and even interact with humans through speech or gestures.

The basis of any AI system is data, as data are needed for effective training of the system and improving its performance. Data serve as the foundation for analysis, decision-making, and the development of insights and knowledge. AI systems mainly benefit from large amounts of data for pattern recognition, feature extraction, bias reduction, and handling variability and rare events. It is important to note that not all AI systems require massive amounts of data. Some algorithms, such as reinforcement learning, can learn from the interaction with an environment rather than relying solely on pre-collected data. Other techniques, such as transfer learning and data augmentation, can also help leverage smaller datasets more effectively. Nonetheless, in most instances, having a substantial amount of variable data contribute significantly to AI systems' performance and generalization ability.

Data refer to raw facts, observations, measurements, or symbols that represent information. It can take various forms and is typically stored and processed nowadays in digital formats for AI ingestion. Data can be classified as structured, semi-structured, or unstructured based on organization and format. As the name implies, structured data refer to data that has a well-defined organization and format. It is typically organized in relational databases composed of rows and columns. Its structure is predefined, making it highly organized and easily searchable. This data type is often represented using standardized formats such as Comma-Separated Values (CSV), Structured Query Language (SQL) databases, or spreadsheets. Some advantages of structured data include easy storage and retrieval, as structured data can be stored in databases with predefined schemas, allowing for efficient storage and retrieval operations; effective analysis that enables straightforward querying, filtering, and aggregation, making it suitable for various analytical tasks; and ease of integration and interoperability due to well-defined schemas that enable smooth integration between different systems and easy data exchange. However, this data type has the significant drawback of being extremely rigid. Structured data require a predefined schema or data model, which can be inflexible in certain situations. If the structure needs to change or accommodate new data attributes, it often requires modifying the schema, which can be time-consuming and disruptive. Other disadvantages of this data type include limited representation, which means the struggle to represent complex relationships or hierarchical structures within the structure; the cost, both in human and computational resources, to maintain the data integrity in structured databases; a lack of efficient scalability of the data; and a lack of contextual information, such as annotations, metadata, or unstructured text.

On the other hand, unstructured data refer to data that does not have a specific organization or predefined structure. It is typically in a raw or natural form and lack a standardized format or schema. Therefore, unstructured data does not fit into traditional rows and columns like structured data. Unstructured data can include various types of information, such as text documents, images, audio files, video files, emails, social media posts, and sensor data streams. It often requires advanced techniques, such as Natural Language Processing (NLP), computer vision, or audio analysis, to extract meaningful insights and patterns.

For this reason, it offers a wealth of information at the expense of a lack of organization, making it challenging to index, search, and analyze without preprocessing it. Additionally, it has high dimensionality and is subjective, thus needing context sensitivity. However, most data created in the real world is unstructured, so for AI purposes, this type of data drives most algorithms.

Finally, semi-structured data refer to data with some organization, but not a fully rigid structure. Semi-structured data may contain tags, labels, or other markers that provide some level of organization or hierarchy. However, the data elements within the format do not necessarily follow a consistent schema across all instances. Therefore, this data type often uses flexible formats like eXtensible Markup Language (XML) or JavaScript Object Notation (JSON). It allows for structure variability and additional attributes or elements to be included as needed. Semi-structured data are commonly found in web data, log files, configuration files, and certain document formats.

Regardless of the data type used for AI creation, its quality, completeness, and processing impact the performance of the algorithms. A common adagio when creating AI models is 'garbage in, garbage out.' Thus, data representation, variability, and fairness are paramount for the development of clinically sound and useful models.

D.2 MACHINE LEARNING

Machine Learning (ML) refers to a specific subset of AI that focuses on developing algorithms and models that enable computers to learn and make predictions or decisions based on data without being explicitly programmed for every possible scenario. In other words, it is a field that aims to create systems that can automatically learn and improve from experience or examples. ML algorithms are designed to identify patterns and relationships within large amounts of data, allowing the system to generalize, make predictions, or take actions on new, unseen data.

For these reasons, broadly speaking, ML is used for five main tasks: automation, classification, regression, clustering, and prediction. Automation refers to the process of automating steps and tasks. It involves using computational systems to perform jobs that typically require manual intervention or human effort due to their repetitiveness, complexity, or attention to detail. By automating processes, ML tools and frameworks enable faster development, deployment, and scaling of products, help reduce manual effort, improve efficiency, enhance reproducibility, and enable humans to focus on higher-level tasks. The classification consists of assigning predefined categories or labels to input data based on its characteristics. ML algorithms can learn from labeled examples and then classify new, unlabeled data points into the appropriate categories.

Similarly, regression involves estimating a relationship between continuous numerical variables. ML algorithms can learn from input–output pairs to make predictions on new data. Finally, clustering is the task of grouping similar data points together based on their inherent patterns or similarities. In contrast to classification, which uses predefined labels, ML algorithms can automatically identify clusters or subgroups in data without predefined categories. These three techniques converge into prediction, which is by far the most common use of ML in clinical medicine. In the context of ML, prediction refers to the process of using trained models to estimate or forecast an unknown or future outcome based on input data. It involves making informed guesses or projections about the value of a target variable or the likelihood of a certain event occurring. As alluded to above, ML algorithms can be categorized into supervised learning (where the algorithm learns from labeled data), unsupervised learning (where the algorithm discovers patterns in unlabeled data), and reinforcement learning (where the algorithm learns through trial-and-error interactions with an environment). The accuracy and reliability of predictions depend on various factors, including the quality and representativeness of the training data, the choice of the ML algorithm, the suitability of the features, and the model's generalization capability. Thus, continuous monitoring and updating of the model are essential to maintain accurate predictions as new data becomes available.

D.3 DEEP LEARNING

Deep Learning is a subfield of ML that focuses on training Artificial Neural Networks (ANN) with multiple layers, hence the term 'deep,' to learn and make predictions or decisions based on large amounts of data. Deep Learning is inspired by the structure and function of the human brain, particularly the optical cortex, where information processing occurs through interconnected neurons. Deep Learning algorithms aim to automatically learn hierarchical representations of data by iteratively transforming the input through multiple layers of interconnected ANNs. Each layer extracts increasingly complex and abstract features from the input data, enabling the model to learn intricate patterns and representations.

In the context of Deep Learning, a neuron refers to an artificial neuron or a computational unit that mimics the functionality of biological neurons found in the human brain. These neurons are the basic building blocks of ANNs and are mathematical functions that receive one or more inputs, apply a nonlinear transformation or activation function to these inputs, and produce an output. They serve as information-processing units in Deep Learning models, enabling the network to learn and make predictions based on input data.

The key characteristics of Deep Learning include interconnected nodes or neurons organized into layers. These networks have an input layer, one or more hidden layers, and an output layer. Each neuron takes inputs, applies a mathematical transformation, and passes the result to the neurons in the subsequent layer in an 'all or nothing' approach. This means that if the mathematical activation function in the neuron surpasses a certain point, the information is propagated to the next neuron. Otherwise, it is not. The power of this is that in contrast to traditional Machine Learning approaches, where feature engineering is required to extract relevant features from the data, Deep Learning models can automatically learn representations of data from raw input. Deep Learning models learn hierarchical representations, where lower layers capture simple features (e.g. edges or textures) and higher layers capture more complex and abstract features (e.g. object shapes or semantic concepts).

However, this complex behavior makes Deep Learning models act as 'black boxes' where the steps or decisions required to devise a solution to the task get diluted into many interconnections. Furthermore, Deep Learning models are trained using an optimization algorithm called backpropagation. It involves computing the gradient of the model's error with respect to its parameters and adjusting the parameters to minimize the error. This process is repeated over multiple iterations or epochs until the model converges to an optimal solution. This backpropagation training further complicates the reproducibility of ANNs.

Nonetheless, Deep Learning has shown exceptional performance in various domains, including computer vision, Natural Language Processing, speech recognition, recommendation systems, etc. It has achieved state-of-the-art results in tasks such as image classification, object detection, machine translation, sentiment analysis, and voice recognition. Many of the most advanced AI/ML models nowadays use ANNs in one way or another, or even multiple stacked systems.

D.4. NATURAL LANGUAGE PROCESSING AND CHATGPT

Natural Language Processing (NLP) is a subfield of AI and computational linguistics focusing on the interaction between computers and human language. NLP enables computers to understand, interpret, manipulate, and generate human language in a meaningful and useful way. It involves applying various techniques, algorithms, and models to process and analyze Natural Language data, including written text, spoken language, or any form of human communication. Its ultimate goal is to bridge the gap between human language and machine understanding.

Natural Language (NL), or the language used by humans for everyday communication, is characterized by its complexity, diversity, and flexibility, as it is the primary means through which people express their thoughts, ideas, emotions, and intentions. It encompasses various forms of communication, including spoken language, written text, gestures, and facial expressions. Some of its main characteristics include its flexibility and creativity since it allows people to express an infinite number of ideas and concepts using a finite set of words and grammar rules. Thus, it enables speakers to create new sentences and convey complex meanings by combining words and structures in novel ways. Furthermore, NL often contains ambiguities, where a word, phrase, or sentence can have multiple interpretations or meanings depending on the context. Therefore, context sensitivity, where the meaning of a sentence or utterance can depend heavily on the surrounding context, including the preceding discourse, shared knowledge, and situational cues, is of paramount importance when deciphering NL. Other challenges include the significant variability in terms of vocabulary, grammar, dialects, accents, and cultural influences within NL; its pragmatic aspects, such as implied meanings, sarcasm, irony, politeness, and indirect speech acts; and its evolution over time. For all these reasons, understanding and processing NL is a complex task for machines due to its inherent nuances and complexities. However, by effectively handling it, computers can engage in tasks such as text analysis, sentiment analysis, language translation, chatbots, virtual assistants, information retrieval, question answering, and many other applications that involve human–computer interaction and communication.

Recently, NLP has found valuable applications in the medical field, transforming how healthcare professionals access, analyze, and utilize medical information. NLP techniques can be used to extract information from medical documents, including Electronic Health Records (EHRs), clinical notes, and medical literature. NLP can automatically identify and extract relevant data such as patient demographics, diagnoses, medications, procedures, and laboratory results, thereby helping streamline data entry, improve accuracy, and enable efficient information retrieval. NLP-powered systems can also help analyze medical literature, research papers, and clinical guidelines to provide evidence-based decision support to healthcare practitioners. NLP algorithms can extract relevant information and present it in a concise manner, helping clinicians make informed decisions about diagnosis, treatment options, drug interactions, and patient care plans. NLP algorithms can assist in the automated coding of medical procedures, diagnoses, and treatments for billing purposes. These techniques can also be applied to social media data, patient forums, and other online sources to monitor public health trends, track disease outbreaks, and capture patient experiences and sentiments to aid in the early detection of public health risks and enable timely intervention. Finally, NLP models can understand and interpret clinical text, enabling the extraction of structured information from unstructured clinical notes. This includes tasks such as named entity recognition (e.g. identifying medical terms, diseases, medications), relation extraction (e.g. identifying drug–disease relationships), and sentiment analysis (e.g. determining patient sentiments and attitudes in clinical narratives).

The most powerful NLP tool designed and deployed to date is ChatGPT. ChatGPT is a language model developed by OpenAI (San Francisco, CA). It is based on the Generative Pre-trained Transformer (GPT) architecture, which is a Deep Learning model designed to generate human-like text by leveraging the power of transformers, a type of neural network architecture. Transformers' innovation is that they can capture long-range language dependencies by utilizing self-attention mechanisms. This allows the model to focus on different parts of the input text and understand the relationships between words, resulting in more coherent and contextually appropriate text response generation. These networks have demonstrated the ability to generate high-quality and contextually relevant text output, making them valuable tools for various language understanding and generation applications. As such, ChatGPT is designed to engage in conversational interactions and provide human-like responses to user inputs in a conversational manner.

ChatGPT was trained on a diverse range of internet text, allowing it to acquire knowledge about various topics and exhibit a broad understanding of language. It can generate coherent and contextually relevant responses, making it suitable for chat-based applications. The model has also been fine-tuned using reinforcement learning from human feedback, where human AI trainers provide rankings and ratings for different model-generated responses. This process helps improve the model's performance and ensures that it produces more desirable and reliable outputs. It is expected that the robustness and capabilities that ChatGPT exhibits will make a huge impact on NLP tasks in the medical field in the upcoming years. However, it's important to note that while ChatGPT can generate coherent and contextually relevant responses, it may sometimes produce incorrect or nonsensical answers. Initially, it has not possessed real-time knowledge or access to the latest information beyond its training cutoff, which was September 2021. While these shortcomings can be fixed in future versions, care should be taken when using this technology for critical tasks or situations requiring accurate and reliable information.

Index

6-minute walking test (6MWT), 65

A

ABI, see Ankle brachial index (ABI)
AI, see Artificial intelligence (AI)
American Diabetes Association 2020 guidelines, 13
Anatomic imaging, 4, 125, 182
    color correction strategy, 42–44
    telehealth/telemedicine (TM), 40–42
    tissue composition, 35–36
    using Digital Single-Lens Reflex (DSLR) camera, 28, 39
    using smartphone, 28
    wound healing trajectory assessment, 36–38
    wound size measurement (see Wound size measurement)
Ankle brachial index (ABI), 15–17, 64, 65, 75, 76
Artificial intelligence (AI), 5, 122–123
    ChatGPT, 199–200
    data augmentation, 193
    deep learning, 196–197
    definition, 193
    machine learning (ML), 195–196
    natural language processing (NLP), 198–199
    reinforcement learning, 193
    semi-structured data, 195
    structured data, 194
    transfer learning, 193
    unstructured data, 194

B

Bacterial fluorescence imaging
    bacterial pigments, 78, 79
    bioburden, 82, 83
    challenges, 83
    debridement, 79, 80
    image interpretation
        characteristic fluorescence signals, 84, 86
        confounding phenomena, 86, 88
        green fluorescence, 84, 85
        Moleculight® imager, 84
        red fluorescence, 85, 87
    infection detection, 78–79
    point-of-care settings, 78
    sampling, 79, 80
    treatment selection, 79–83
Biospectroscopy, 52, 55

C

Cell tissue products (CTP) grafting/applications, 79, 81–83
Charcot arthropathy, 71, 72
ChatGPT, 199–200
Chronic wound categories, 1–2
Classification of diabetic foot ulceration (DFU), 2
Clinical signs and symptoms (CSS), 24, 81
Clock method, 30
CMOS camera
    design considerations, 188–190
    limit of detection (LOD), 183–184

limit of quantification (LOQ), 183–184
noise, 184–186
pixel sampling area, 186
remote PPG signal, 187
sampling frequency, 186
sampling tissue heterogeneities, 186–187
Coefficient of extinction, 172
Comprehensive patient assessment, 3
Confocal microscopy/confocal laser scanning microscopy (CLSM), 89
Constructive and destructive interference, 170–172
Contact methods, 29, 32–34
Contact planimetric methods, 32
Critical limb ischemia (CLI), 20, 64, 68

D

Data augmentation, 193
Deep learning, 196–197
Deep tissue infections, 73
Delphi consensus method, 80, 81
Diabetic neuropathy, 2, 8, 9, 69–70
Digital imaging methods, 31–33, 39
Direct contact methods, 34
Distal symmetric neuropathy, 8, 9
Dynamic thermal analysis (DTA), 66
Dynamic thermography (DT), 65–67, 76

E

Electronic health record (EHR) system, 128–130
Endovascular therapy (EVT), 64–65
Exercise-induced Temperature Change (eTC), 65
Exogenous fluorescence imaging, 76–77

F

Fast Healthcare Interoperability Resources (FHIR), 129
Flap perfusion, 66–68
Fluorescence angiography (FA), 76
Functional ischemia, 11

G

Greatest length and width method, 30

H

HSI/MSI, *see* Hyperspectral and multispectral imaging (HSI/MSI)
Hydrostatic modulation test, 66
Hyperbaric oxygen therapy, 55, 56
Hyperspectral and multispectral imaging (HSI/MSI)
  in biomedical applications, 53
  clinical applications, 55–56
  deoxyhemoglobin (RHb) measurement, 54
  diabetic foot ulcer (DFU) development prediction, 55
  epidermal thickness, 56–57
  hyperbaric oxygen therapy, 55, 56
  near-infrared (NIR) method, 57, 58
  oxygen content visualization, 53, 54
  oxygen saturation (SO_2) maps, 53
  oxyhemoglobin (HbO_2) measurement, 54
  post-debridement imaging, 58–59
  visible light method, 57, 58
Hypoxia, 20, 24, 48

I

Imaging photoplethysmography (iPPG), *see* Remote photoplethysmography (rPPG)
Incident dark field (IDF) technique, 89
Indocyanine green angiography (ICGA), 23, 76
Inflammation detection, 70–72
Infrafemoral endovascular procedures, 68
Infrared thermography (IRT)
  absolute *vs.* relative temperature gradients, 61, 63, 74
  angioplasty assessment, 68, 69
  clinical considerations, 75
  clinical thermographic studies, 60, 62
  core *vs.* skin temperature, 74
  dynamic thermography (DT), 65–66

endothelial dysfunction, 69
endovascular therapy (EVT), 64–65
flap perfusion, 66–68
infection detection, 71–73
inflammation detection, 70–72
infrared (IR) detectors, 60
microvascular dysfunction, 69
thermal ABI (tABI), 64
thermal asymmetry, 61, 63
thermal camera applications, 59
thermal imaging sensors, 59–60
uncooled detectors, 60
viewing angle, 75
wound bed *vs.* peri-wound, 74
Interoperability, 128–130, 194
IRT, *see* Infrared thermography (IRT)
Ischemia, 48, 60, 123

K

Kelvin scale, 43
Kruithof diagram, 43, 44
Kundin® gauge, 33

L

Large fiber tests, 14–15
Laser antiseptics
   photodynamic therapy (PDT)
      antibacterial photodynamic therapy (aPDT), 106
      antibacterial properties, 106
      antimicrobial properties, 106
      application of, 105
      mechanism of action, 106–107
      photosensitizers, 107–109
   phototherapy, 109–111
   UV-C, 105
Laser debridement, 103–104
Laser Doppler flowmetry (LDF), 50–51, 183
Laser speckle (perfusion/contrast) imaging (LSPI/LSCI), 51
Laser systemic therapy, 111
Laser triangulation method, 34
Lifetime risk, 1
LifeViz®, 35
Light absorption in biological tissues
   absorption coefficient, 155
   definition, 155
   energy dissipation mechanisms
      chemical reactions, 162–163
      fluorescence, 160–162
      Jablonski diagram, 160
      photoacoustics, 162
   tissue chromophores
      β–carotene, 159
      bilirubin, 159
      cytochrome c oxidase, 158–159
      epidermal melanin, 155–157
      fat, 157–158
      hemoglobin, 156
      water, 157
Light propagation in tissue, 172–173
Light properties
   coherence, 152
   monochromatic *vs.* polychromatic, 152–153
   polarization, 153–155
   spectral ranges, 150–152
Light refraction in biological tissues
   ray bending, 163
   ray reflection, 164
   refractive index, 163
Light scattering in biological tissues
   definition, 165
   elastic scattering
      lifetime, 165
      Lorenz–Mie–Debye solution, 165
      Lorenz–Mie solution, 165
      Mie scattering, 166
      origins of, 168
      Rayleigh scattering, 166
      reduced coefficient of scattering, 167–168
      scattering anisotropy, 166–167
   inelastic (Raman) scattering, 168–170
   laser Doppler, 170
   scattering coefficient, 165
   scattering cross-section, 165
Limb thermography, 65
Linear healing rate, 37
Lower-limb amputation, 1
Low-level laser therapy (LLLT), 5, 103, 112, 113, 115–116
Low-level local biostimulation

clinical data, 112–113
combined methods, 113–114
Cytochrome c Oxidase (COX), 114
light-stimulated wound-healing process, 114
open questions, 115–116
phototherapy, 114, 115

M

Machine learning (ML), 195–196
Macrocirculation, 18, 19
Macrovascular disease, 10
    ankle brachial index (ABI), 15–17
    Doppler ultrasound probe, 18
    toe brachial index (TBI), 17–18
Management of diabetic foot ulceration (DFU), 2–3
Mathematical model method, 30–31
Mean optical path (MOP), 174, 175
Mean sampling depth (MD), 175
Medial calcinosis, 10, 16, 17
Metabolic fluorescence imaging, 90–91
Microbiological studies, 24–25
Microcirculation
    fluorescent imaging, 22–24
    skin perfusion pressure (SPP), 21–22
    transcutaneous oxygen pressures ($TcPO_2$), 19–21
Microvascular disease, 10–11
Multimodal devices, 126–127
Muntean's method, 67

N

Nanomedicine, 123–124
Natural language processing (NLP), 198–199
Near-infrared spectroscopy (NIRS), 52–53
Neuropathy examination, 13–15
Non-contact optical methods, 34–35
Nyquist–Shannon sampling theorem, 187

O

Optical coherence tomography (OCT), 152, 171–173
Optical density (OD), 176

Optical metabolic imaging (OMI), 90–91
Orthogonal polarization spectral (OPS) imaging, 87
Oxygen saturation (SO_2) maps, 53

P

PAD, *see* Peripheral arterial disease (PAD)
Patient education and self-management, 126
PDT, *see* Photodynamic therapy (PDT)
Perfusion targeting techniques
    blood oxygen saturation, 50
    exogenous fluorescence imaging, 76–77
    hyperspectral and multispectral imaging (HSI/MSI)
        in biomedical applications, 53
        clinical applications, 55–56
        deoxyhemoglobin (RHb) measurement, 54
        diabetic foot ulcer (DFU) prediction, 55
        epidermal thickness, 56–57
        hyperbaric oxygen therapy, 55, 56
        near-infrared (NIR) method, 57, 58
        oxygen content visualization, 53, 54
        oxygen saturation (SO_2) maps, 53
        oxyhemoglobin (HbO_2) measurement, 54
        post-debridement imaging, 58–59
        visible light method, 57, 58
    infrared thermography (IRT)
        absolute *vs.* relative temperature gradients, 61, 63, 74
        angioplasty assessment, 68, 69
        clinical considerations, 75
        clinical thermographic studies, 60, 62
        core *vs.* skin temperature, 74
        dynamic thermography (DT), 65–66
        endothelial dysfunction, 69
        endovascular therapy (EVT), 64–65
        flap perfusion, 66–68
        infection detection, 71–73
        inflammation detection, 70–72
        infrared (IR) detectors, 60
        microvascular dysfunction, 69

thermal ABI (tABI), 64
thermal asymmetry, 61, 63
thermal camera applications, 59
thermal imaging sensors, 59–60
uncooled detectors, 60
viewing angle, 75
wound bed *vs.* peri-wound, 74
laser Doppler flowmetry (LDF), 50–51
laser speckle (perfusion/contrast) imaging (LSPI/LSCI), 51
near-infrared spectroscopy (NIRS), 52–53
phosphorescent long-term oxygen sensors, 77
spatial frequency domain imaging (SFDI), 59
Peripheral arterial disease (PAD), 8
ankle brachial index (ABI), 15–17
cooling challenge, 66
critical limb ischemia (CLI), 64–65
diagnosis of, 60–61
Doppler ultrasound probe, 18
indocyanine green angiography (ICGA), 76
symptom, 10, 69
toe brachial index (TBI), 17–18
Phosphorescent long-term oxygen sensors, 77
Photodynamic therapy (PDT)
antibacterial photodynamic therapy (aPDT), 106
antibacterial properties, 106
antimicrobial properties, 106
application of, 105
mechanism of action, 106–107
photosensitizers, 107–109
Photoplethysmographic (PPG) imaging, 182
Photosensitizers (PS)
nanostructures, 108–109
natural photosensitizers, 108
prodrugs, 109
synthetic dyes, 107
tetrapyrroles, 107–108
Phototherapy, 105, 109–115
Physiological imaging, 4, 5, 125
CMOS camera (*see* CMOS camera)
definition, 182
Planimetric techniques, 29–30

Polyneuropathy, 8–9
Pre-EVT thermal asymmetry index (TAI), 65
PS, *see* Photosensitizers (PS)
Pulse oximeter, 174

Q

Quantum dots (QDs), 124

R

Radiometry, 177–178
Reinforcement learning, 193
Remote monitoring and alerts, 126
Remote photoplethysmography (rPPG), 90
Resolution-sampling depth coordinates, 49
Revascularization methods, 68

S

Sampling depth, 174–176
Semiautomatic segmentation, 31
Semmes–Weinstein Monofilament (SWM) examination, 174
Silhouette Star®, 34
Skin morphology
blood, 139–140
blood vessels, 139–140
dermis, 138
epidermis, 136–137
subcutaneous tissue, 138–139
Skin perfusion pressure (SPP), 21–22
Skin types, 135–136
Small fiber tests, 14
Smartphones, 125–126
Society for Vascular Surgery (SVS), 48
Spatial frequency domain imaging (SFDI), 59
Speckle formation in biotissues, 170–171
Speckle plethysmography (SPG), 31, 35, 89–90
Structured light/active illumination profilometry techniques, 34
Surface reflection discarding techniques, 89
Swift HealX® marker, 44
Swift Medical Patient Connect App, 40

T

TBI, *see* Toe brachial index (TBI)
Telemedicine (TM), 40–42, 125, 130–131
Theranostics, 123–124
Therapeutic windows, 177
Thermal ABI (tABI), 64
Tissue chromophores
  β–carotene, 159
  bilirubin, 159
  cytochrome c oxidase, 158–159
  epidermal melanin, 155–157
  fat, 157–158
  hemoglobin, 156
  water, 157
Tissue composition, 35–36
Tissue reflectance, 176–177
TM, *see* Telemedicine (TM)
Toe brachial index (TBI), 17–18
Transcutaneous pressure of oxygen (TcPO$_2$), 19–21, 50
Transfer learning, 193
Treatment
  implementation, 3
  initial therapy, 102
  invasive surgery, 103
  laser antiseptics
    photodynamic therapy (PDT), 105–109
    phototherapy, 109–111
    UV-C, 105
  laser debridement, 103–104
  laser systemic therapy, 111
  low-level local biostimulation
    clinical data, 112–113
    combined methods, 113–114
    Cytochrome c Oxidase (COX), 114
    light-stimulated wound-healing process, 114
    open questions, 115–116
    phototherapy, 114, 115
  non-traumatic lower limb amputations, 103
  planning, 3
  surgical intervention, 102, 103
  wound cleaning, 103–104
  wound closure, 104

U

U.S. Center for Disease Control, 1

V

Vasculopathy, 9–11
Venous leg ulcers (VLU), 37
Virtual consultations, 125

W

Weight-to-volume method, 34
Wound care management apps, 125
Wound care workflows, 3, 38–40
Wound closure, 36, 38, 104, 112
Wound debridement, 82, 104, 143
Wound diagnosis, 3
Wound healing, 1, 24
  trajectory assessment, 36–38
Wound inflammatory index/temperature index, 71
Wound measurement apps, 125
Wound size measurement
  area measurement
    digital imaging, 31–33
    mathematical model method, 30–31
    planimetric techniques, 29–30
    simple ruler method, 30
    stereophotogrammetry, 31
  geometrical wound measurements, 28
  three-dimensional (3-D) methods, 29
  two-dimensional (2-D) methods, 29
  volume measurements
    contact methods, 32–34
    non-contact optical methods, 34–35
Wound tissue
  adipose tissue, 146
  bone, 146
  corns and calluses, 146–147
  epithelial tissue, 141, 144
  eschar, 142–143
  granulation tissue, 141, 143–144
  muscles, 145
  necrotic (nonviable) tissue, 142
  nerves, 145
  slough, 142, 143
  tendons and ligaments, 146